Managing a Paint Shop

Managing a Paint Shop

fundamentals of leadership and organization

by Robert D. Grear, CMfgE, CM

Published by
The Society of Manufacturing Engineers
in cooperation with the Association
for Finishing Processes of SME
One SME Drive
Dearborn, Michigan 48121

Copyright © 1994 by Society of Manufacturing Engineers

098765432

All rights reserved, including those of translation. This book, or parts thereof, may not be reproduced in any form or by any means, including photocopying, recording, or microfilming, or by any information storage and retrieval system, without permission in writing of the copyright owners.

No liability is assumed by the publisher with respect to the use of information contained herein. While every precaution has been taken in the preparation of this book, the publisher assumes no responsibility for errors or omissions. Publication of any data in this book does not constitute a recommendation or endorsement of any patent, proprietary right, or product that may be involved.

Library of Congress Catalog Card Number: 94-067171
International Standard Book Number: 0-87263-453-1

Additional copies may be obtained by contacting:

Society of Manufacturing Engineers
Customer Service
One SME Drive
Dearborn, Michigan 48121
1-800-733-4763

SME staff who participated in producing this book:

Donald A. Peterson, Senior Editor
Dorothy M. Wylo, Production Secretary
Rosemary K. Csizmadia, Operations Administrator
Sandra J. Suggs, Editorial Assistant
Cheri Willetts, AFP/SME Association Manager
Judy D. Munro, Manager, Graphic Services

Printed in the United States of America

About AFP/SME

The Association for Finishing Processes of SME (AFP/SME) is the principal educational association for those involved with paint and coating applications worldwide. It was founded in 1975 to provide an informational interchange among producers of coatings and pretreatment applicators, product designers, equipment manufacturers, users and coatings operations professionals, those responsible for environmental compliance, and finishing consultants.

The Association's goal is to provide leadership and knowledge concerning the finishing industry. Its objectives are to expand the use of new technologies and meet the rapidly changing needs of its constituents. AFP/SME serves individuals by promoting finishing as a career. It serves companies by promoting technical excellence. The Association also serves the environment by promoting effective management and operational techniques.

In pursuit of its goal, AFP/SME sponsors conferences, clinics, and courses around the country. Topics for these sessions include: industrial painting processes, production painting, systems design, troubleshooting coating defects, liquid and powder coating, radiation curing, plastic finishing, robot finishing, wood finishing, and environmental compliance.

In addition, AFP/SME publishes books, reports, *The Finishing Line* technical quarterly, and pocket guides to provide current technical information on all aspects of industrial finishing.

In joining AFP/SME, an individual also becomes a member of the Society of Manufacturing Engineers and receives *Manufacturing Engineering* magazine. The Society presently comprises over 75,000 manufacturing-oriented members in some 70 countries throughout the world.

For more information, call the AFP/SME Manager at (313) 271-1500 ext. 544.

Preface

The need and value of properly painting and finishing a product have never been more important than they are today. Moreover, there is no indication that their importance is diminishing. If anything, the paint line will contribute more and more to the marketability, quality, cost, design, packaging, and other links in the total manufacturing food chain in the future. Certainly it will impact the environmental aspects of our manufacturing strategies.

One of the greatest problems in my nearly 40-year career has been obtaining qualified management for paint operations. The discipline itself has lacked the professional image and performance found in other areas such as welding, stamping, and machining. Until only recently, there were really no formal degrees available in paint technology and few ways to obtain any form of professional certification. We must change that perception of paint management.

It was my purpose in preparing this book to share insights and experiences from my involvement in paint operations. Paint management is still very much misunderstood by many top-level management people. The needs vary widely; those of a small manual operation differ greatly from those of a more sophisticated automated facility. Some traits and activities, however, are common and essential to the success of all forms of paint operations, large and small. A combination of substance and style can be modified to fit any condition found or need uncovered if applied in a planned and timely manner. Leadership and management in the paint shop make it happen. When present, they turn losers into huge winners.

The content of the book is solely that of the author. It comes from a broad range of experience and association with hundreds of people over a long period. Some of it is a blur that has become so much a part of my life I don't always recall the whos and wheres.

Acknowledgments

Although much of the book is original in both content and application within painting, it would be improper not to pay homage to a few who have been of very special help in putting it together. People like Terry Simonton and Fred Scheetz. These two long-time associates have been my help so many times in accounting and budgetary matters. Then there is Mary Bullwinkel, one of those "girl Fridays," who as secretaries, seldom share in any of the limelight. She just got the work, and it always has to be done correctly and on time, even though we did not get the information to her until the very last moment. Next comes the people on staff at the Society of Manufacturing Engineers. People like Cheri Willetts, Karen Wilhelm, Bob King, and Don Peterson who have to be the glue that holds things together for the people like myself.

A special thanks also goes to Jack Stauffer for his contribution on painter's math and Keven Loop for input into charts and quality activities. Special thanks also goes to Tim McDaniel for his assistance on environmental issues.

Next, I want to recognize Al Jones for his insights into union relations and contractual understanding in both negotiating and operational utilization of labor agreements.

Finally there is my wife, Mickey. She is the best manager I have ever known. No one can get more out of few assets and help reach the goals of the organization better than she. There is also a little friend named Chynna who will lay at my feet for hours with me while I write and put together some plan for the next adventure in painting.

Robert D. Grear, CMfgE, CM

Table of Contents

Preface..vii

Introduction...1

1. Leadership and Management..5
2. Finishing is a Philosophy, an Aid to the Modern Manager...............11
3. Teach Me How to Fish..21
4. Putting Together an Organization..27
5. Poor Management, Entitlement, or Work Ethic Problems................41
6. Staffing the Workplace in the Seniority Game.................................49
7. Seed Corn, Ice Cream, and Turtles...55
8. Business Planning in the Paint Shop..61
9. Budget Planning...77
10. Planning Facilities and Painting Strategies....................................91
11. Monitoring for Cost Control..113
12. How good is Good? ISO 9000, SPC, and Quality Control............125
13. Painters' Math..151
14. Problem Solving and Decision Making..163
15. Maintenance (Not an Expense, a Contributor).............................177
16. Environmental, Health, and Safety Considerations.....................201
17. Supplier Relations/Responsibilities..221
18. Sabotage and Other Negative Communications..........................227
19. Winning the Professional Way..233

Et Cetera..241

Bibliography..259

Index..261

Introduction

Several people have asked me what makes a good paint manager. My response is simple: There probably is no single, absolute formula or answer. There are a number of experiences in my career that can shed some light on just how one measures the success or failure of a good manager.

Some people consider themselves good managers simply on the basis of "getting the numbers out the door" each day, etc. But they may treat their human and physical assets terribly. They may tolerate unsafe working conditions. And their costs may be grossly out of control to the extent that no one has much of a future in their job.

This scenario obviously does not describe a good manager. During my career, I have used three basic guidelines to frame every decision and define each action:

1. Give the product the benefit of doubt. The product makes everything else possible and stands as your signature to the world of your pride in craftsmanship. Continuous improvement is management's responsibility, and must be passed on to everyone having a part in producing the product.
2. Make the safety of your workers your highest priority. There is no bucket of paint, chemical, or product worth an unsafe workplace. Period!
3. Represent yourself and your company with the greatest integrity. Your credibility is perhaps the strongest asset you may ever enjoy in both personal and business relationships.

These basics have brought a great deal of success in my career, and I highly recommend them. They can serve as the foundation of your building a strong career at any level of management. They remain constant over the long term, and they work!

From my experience, I have learned that management embraces a combination of both substance and style. I have never known a successful organization that did not possess that combination to some degree, paint shops included. Paint shops have changed a great deal through the decades: facilities, materials, people, and product are all a part of that change. Although the skills necessary to manage today are quite different from those of the 1950s, the elements of substance and style still define management technique.

The following chapters contain a great deal of substance. There are examples of organizational considerations, strategies, painter's math and accounting, problem-solving, costs and quality control, environmental and safety requirements, maintenance, communication skills, and supplier relationships/responsibilities.

But important as substance is, management style is playing an increasingly critical role in the management equation, due to advances in technology and the demands being put on the people required to support this technology. The day of the "bull of the woods" manager is all but gone. Communication skills, which have always been important, emerge as a prime requisite for management in the coming decades. The era of "The beatings will continue until peace and harmony once again exist" is yielding to one of participation and cooperation. A positive style and communicative effort is found in all winning organizations. Browbeating will never produce long-term winners.

It has also been my observation that finishing is like an iceberg. Only a small part of the total can been seen above the water by the layman. The part that sinks the ship is below the surface, and the effective manager had better know what that represents in steering a course safely around it.

Good managers need to know not only their own specific area, but also how their operations impact other departments and divisions in the company, and vice versa. For example, the accounting group of your firm may be outdated when it comes to modern technology. It may still be using systems designed around direct labor in manual application paint lines. Robots or other forms of automation that eliminate direct labor may spawn totally inaccurate budget data now if the data is compiled against direct labor criteria. It would be apples times oranges, and that doesn't compute. You may not be tracking costs properly.

All businesses are some recipe or combination of what I call the "M & Ms," parts of a total that all start with the letter "M." Materials, manpower, methods, machinery, money, to mention just a few. They are always present and will always be found in various proportions. To me, these constitute a great deal of the unseen part of the iceberg. A good manager, like a good cook, knows how to put them together in the proper perspective, add time to market, and create a winning product instead of a generic product.

I have selected each part of this book for its importance to the painting operations manager. Because shop operations are so interdependent, it is difficult sometimes to break off discussion on one subject and go on to the next. Topics often spread across lines and chapters. No chapter is expected to contain all that can be written on a given subject. Nor does chapter order even reflect any certain priority of importance depending upon circumstances.

Several roads lead to the town square. I will attempt to show you at least one way to get there, or at least help you recognize the various signs for mapping out a route to satisfy your needs and goals as a paint-shop manager.

I could probably have made a living doing a number of things. The finishing discipline has been a wonderful career for me. It has been satisfying; it has afforded opportunities unheard of as a child in rural Ohio coming out of the depression. It has allowed me to meet and associate with some of the brightest and finest people to be found in the world, and get paid for it.

If this book attracts or retains just a few bright young people to the finishing discipline as career professionals, the effort will have been well served. The painting discipline is now a mature one. To carry on, we need good talent with the integrity to take us through the challenges ahead. We must compete on a world-class basis and meet increasingly tough environmental issues. If we don't, our manufacturing strength will cease to exist.

I make no claims to knowing it all. I do want to share my expertise and experience with anyone who is receptive. Hardly any manufactured product can be shipped without first being painted or finished in some form. The bottom line reflects how well, safe, and cost effectively we do.

CHAPTER 1

Leadership and Management

Vision is central to leadership. True leaders consistently exhibit the unique capability of anticipating needs and defining solutions.

What are the attributes of a well run organization? The symptoms of one in decline? What makes one company's finishing department excel and another's barely get by?

Simply put, the quality of leadership and the effectiveness of management. Both are vital to the success of any enterprise. In large companies, they are likely provided by more than one person, while in small shops, they could be embodied in the same person. But whatever the scenario, that which constitutes leadership and that which makes up management are significantly different entities.

Expectation and achievement

Leadership has been defined in many ways by many people through the years. The late coach Vince Lombardi of the Green Bay Packers once said, "To me, a leader is a visionary that energizes others. This definition of leadership has two key dimensions: (a) creating a vision of the future, and (b) inspiring people to make the vision reality...."

Leaders tend to have a quality of being able to anticipate or define needs and wants better than the average person. They have the confidence to form a plan and see the potential end result clearly and quickly. Leaders also have a way of focusing energies and moving ahead in a resolute manner, picking up support along the way. They move organizations in new directions.

Leaders generally are great communicators and generate feelings and meaning to others around them. Work is not just a job to a leader; it has value and adds meaning to the effort. Leaders inspire emotions and invoke values in an organization.

Another characteristic of leaders is that they take risks. But they usually do not view the level of risk as severely as those around them. The leader's vision or seeming "x-ray" capability of embracing and comprehending small details of plans is the "calculated" element not recognized by others that lessens the hazard of risks taken. Even when the risk is great — but necessary — the leader's skills and knowledge will temper the jeopardy.

Obviously, no organization can exist long without leadership. Yet, all too often it is mistaken for a more common concept called management. Where leadership involves a range of personal, somewhat intangible attributes, the characteristics of what we call management are more definable and measurable.

The traditional functions of management are planning, directing, controlling, organizing, and staffing. In fact, a whole industry has been founded on defining, teaching, and implementing a wide variety of management concepts.

Behaviorists see management as a complex of interpersonal relationships, relationships that serve as the management basis for the use of psychology. There are those who see management as an example of institutional and cultural aspects of society. Still others view management as decision making in all aspects of an organization.

There are those who feel management is an exercise in mathematics, logic, and ever-increasing use of models. Today we have psychologists, sociologists, anthropologists, sociometricists, economists, mathematicians, physicists, biologists, political scientists, business administration scholars, and even practicing managers attempting to get on this interesting and sometimes profitable bandwagon.

This is often upsetting to the average practitioner, who both sees and desires the benefits that can come from effective management. Many of the leaders of management movements have academic or other interests feeding their zeal to carve out some original and distinct approach to management. And to protect their interests, many promote these causes at the expense of practitioners "in the trenches" whose hands-on knowledge was gained by doing.

Putting management to work through leadership

Numerous schools of management have come and gone over the years — the empirical, human behavior, social system, decision theory, and mathematical forms — all differing widely, but all having similar goals and dealing essentially with the same world of organizational management.

Management, with all academic mist and mystique removed, is still just getting things done through and with people.

The intent here is not to introduce new theories of management. Instead, it is to show how to use the best parts of several successful and proven techniques in the operation of a paint area, which is an integral part of the total manufacturing mix.

It is safe to say that management's first function is to run the business and put economic performance first. Its effectiveness is measured in terms of the economic results it produces. There are noneconomic results, of course, such as worker morale and contributions to the welfare of the community, culture, and other areas of society, but if management does not deliver the goods and services demanded by the market at a price the customer is willing to pay, it has failed. It has not added value from the economic resources entrusted to it, and, by extension, all other noneconomic areas will feel the impact of this failure.

Management is not and can not be an exact science. It embraces distinct professional attributes and some scientific characteristics. And while some management skills seem to emerge from innate ability, most of the skills can be learned by anyone having the desire. Most modern methods seek to reduce the role of the "intuitive" manager to a system of acquired knowledge, principles, and analysis. It is almost expected that managers today have the self-discipline and dedication to high standards that define a true professional. In reality, too often this has been the missing ingredient in the formula for effective paint shop management.

Management — its level of competence, integrity, and performance — ultimately plays the deciding role in determining the futures of entities at all levels — families, companies, nations, and the world. The responsibility is great and demands continue to increase.

Yet despite its importance, management is probably one of the least understood functions of a company. This is especially true in the paint shop. The front office's knowledge of paint operations is typically found wanting.

And, in the minds of the paint people, the front office appears blind and indifferent. (In fact, a stigma of low-blue-collar non-professionalism attached to paint people today likely stems from the type of painters populating our shops 40 years ago. Though the quality of paint people has changed, the stigma persists.)

The human element

There are those who used to believe economic performance could be made by putting together human and material resources in an almost mechanical order. However, this does not work. It leaves out a critical part of the formula; there must be some transformation of the resources, and this doesn't come from an inanimate object. The only resource capable of *real* enlargement is the human one. All others operate under the laws of mechanics. They may be manipulated for better or worse, but they can never have an output greater than the sums of their inputs. Only people can grow and develop.

In the past, the rank-and-file worker was viewed as a person who functioned only as directed, without sharing responsibility or having a voice in the decisions of his or her work or that of others. In practice, they were looked upon much like any other material resource.

The product of this practice has been painfully wrought. Failing to recognize that from today's rank-and-file come tomorrow's managers, upper management neglected to cultivate potential managers from the shop floor. Only recently has it been acknowledged that developing managers is really what turns a collection of resources into a viable enterprise, and that managing managers is itself a crucial function of management.

A final function of management is to manage work. This means providing the tools and environment sufficient for human beings to work productively in a safe and effective manner. Included in this is the factor of time. Every thought, function, decision, and action must be considered in the context of both the present and long-term survival of the organization. When these are not balanced and in harmony, the resources will be endangered, damaged, or destroyed.

It is management, then, that takes an idea or plan, puts it into action, and provides the control for all phases of its implementation. In tandem, a manager has the responsibility of coordinating and controlling the physical, technological, and human resources of the organization such as property, equipment, materials, processes, and workers.

Strategist or tactician?

While leaders focus on long-range goals, managers are expected to attain short-term objectives and goals. Managers communicate policies, procedures, and other directives. It is their responsibility to obtain the optimum performance from existing facilities or processes. This includes maximum focus on continuous improvement.

By nature of their organizational authority level, managers are not normally given the financial authority to make great directional changes. These higher-level changes are generally subject to all forms of investigation and scrutiny before potential approval. Mid-level managers, predictably, tend to act cautiously and feel most comfortable with a minimum of change from outside influences. Hence, changes they make tend to be incremental, not fundamental.

Managers not only communicate policies, they are expected to enforce them. It is here that they quite often find conflict, arising from not agreeing with policies and not wanting to be potentially embroiled in conflict. This is particularly common in the local workplace, such as a paint shop, where managers have come from the work force and are former friends and associates.

Fortunately, these traditional methods of management are disappearing. The new wave is toward approaching people and their assignments so they are empowered to apply their knowledge to increase productivity and improve job satisfaction and customer service. Implicit in this trend is the need for managers to develop techniques that inspire people.

All too often people are failing to lead at times when leadership is critical. They are just managing, loading only one side of the fulcrum. This begs the question: Where is the balance? Look around and see how many winning people are called managers. We have world leaders, community leaders, church leaders, business leaders, political leaders, etc. None are called managers, but the requisite is implied. Each at some time has had to play both roles, but it is the leadership qualities that create something for others to build upon. An individual is limited to just how much he or she can physically and mentally perform. But motivation is the ally of the effective manager-cum-leader; by inspiring others, his or her capability is multiplied through each worker, creating a synergy that produces a whole greater than the sum of the parts. Modern "management" blurs the traditional roles of leaders and managers and forces the leader quality to be more evident.

Today's manager (or leader) moves back and forth across both lines more and more as organizations take on modern methods. More team and consensus decision making takes place in the contemporary workplace than it did in former times. This is permitting input into planning upward into the organization rather than everything flowing one way down to people. This can be a difficult transition, whether it takes place in the executive suite or at the floor level of a plant. The roles can become easily blurred. Executives and managers alike often fear their responsibility or authority has been diluted, while workers often take pride in being asked to participate. But whether on the top floor or the shop floor, the effort must be made in an open and honest manner. People are quick to sense when shared responsibilities are just being given the "buzz word" treatment. This can lead to total destruction of communications and morale.

The onus of accountability

Being a leader is perceived to carry with it a lot of advantages in recognition, pay, and other forms of reward. To a degree, this is true. Certainly the leader has the ability to influence events within an operation. This reward brings on a much greater level of responsibility, however. It isn't always easy. Leadership, or a lack of it, affects people, plants, companies, and cities. Leaders and managers are highly visible, not part of a group, and will never be eight-hour-per-day persons.

That's part of the price that goes with either title. It takes sacrifice and a lot of effort to continually acquire and apply knowledge.

Probably the greatest lesson a manager and leader will learn is that the more responsibility one gains, the more one becomes a servant to all he or she is accountable for. The manager/leader must satisfy both the needs and wants of those making up the organization, contrary to what might be conventional belief. Leadership and management are functions of serving others.

CHAPTER 2

Finishing is a Philosophy, an Aid to the Modern Manager

Management's task is to nurture the human resource through training and guidance. It must provide the direction and articulate the goals under which the organization is to function.

Certain considerations seem always to distill upward when managers embark on strategy definition in the corporate cosmos. And the paint shop is not immune. For instance, what do we want to be as a finishing operation? Who are we and what is our potential contribution to our company, our fellow workers, our community, and our industry?

The strategic view

Finishing, viewed as a philosophy, is a strategy, just as any other discipline in daily business. Paint, chemicals, and equipment have no capacity for adding intelligent value to anything. They are unable by themselves to be good in one area and bad in another. Human intelligence is the catalyst that binds the process with the product to add the competitive advantage of value.

The flexibility inherent in the finishing function accommodates a broad range of manufacturing activities. Various substrates can be utilized and differing welding, bonding, and assembly techniques can be applied. The finishing function also permits changes in styling concepts and material handling methods and improves quality, safety, cost issues, and environmental compatibility. Finishing represents a lot of contribution to product success, when utilized effectively, and is critical to product acceptance. In truth, few products can be produced without some level of finishing.

But for all its importance, painting is like an iceberg. Only a small part is visible to management and laymen alike. It is the submerged portion that can severely damage product performance and acceptance or cause it to outpace competition hands down. It is important, then, to understand that finishing begins a long time before any product ever enters the finishing area.

Finishing should begin with product design. It is at that point that several vital issues must be resolved. What is the product going to look like? How will it be manufactured? What type of substrate will be required? What type of lubricants or contaminants will be used during manufacture? How can it be transported and presented for cleaning and finishing? What is its weight, size, and ability to accept chemicals and heat? Can it be grounded? How good is good? Will the product be run in high volumes or batches of low volumes? The best process to meet market demands will in large part be determined by answers to these and similar questions.

A good case in point is that of a major truck builder with whom I worked who was looking to redesign his product's front portion. Prompting the redesign were three issues needing resolution: a problem of corrosion on trucks, new fuel mileage regulations, and demands for easier servicing and repair in the field.

The front ends of these vehicles had traditionally been produced with up to 10 large, separately manufactured parts which were eventually assembled as a front end. These parts also incorporated numerous brackets and subassembly parts needing much welding. In addition, a high level of material handling was involved which led to dings and other product damage requiring repair before finishing and subsequent assembly.

Discussions to meet these needs led us to decide on a one-piece fiberglass front end. It would be designed to tilt forward for easier access and servicing of vehicles.

We found the fiberglass could be made stronger than the existing metal parts and the new hood assembly was lighter in weight to help improve fuel consumption. And, importantly, this substrate would not rust.

There were other considerations, however. The new design would eliminate almost 40% of all the welding operations internally, and change the routing, receiving, and carriers. And in finishing operations, it would place the unit in the paint line at another point.

Fortunately, years before, we had held out for power and free conveyors in the paint shop instead of traditional monorail lines. With this system, it was possible to run metal parts through pretreatment systems and prime them before introduction of the new one-piece hoods. It always had been critical to control color match between bodies and front-end pieces in the original metal parts because they were painted in separate small parts lines. Now it was possible to eliminate some small parts booths and schedule the large one-piece hoods right along with the cab for each. This assured a color match since the same paint and painters were now painting an entire unit as one.

One of the attributes of designing with plastic is that you can easily change the lines and shapes of the product many times during design as opposed to metal products. Design engineers had to learn this, but the result was a better styled and upgraded appearance of an older model, resulting in a lot more life from the initial tooling investment.

From this one design action, the paint shop got a new substrate to prepare, prime, and finish paint. It had new material-handling and scheduling considerations, different paint mix needs, and a new set of grounding considerations to continue the use of electrostatic guns. The paint shop gained better color control, produced less scrap and fewer color changes, experienced no corrosion, and even cut costs. It was not necessary to prime or paint the back sides of the fiberglass product since corrosion was not a consideration. From this came labor reductions, increased capacity in some areas because of only painting one side, fewer volatile organic compounds (VOCs), and reduced maintenance/disposal costs.

The customer received a better quality truck that serviced easier and looked better. The new design helped improve fuel efficiency and lower operating costs while improving environmental conditions. In the end, the customer got all this at a very competitive price.

Whether the reason for change is internal or external, the important thing to remember is that finishing begins a long time before products come to the paint shop. By planning finishing processes concurrent with product design, and identifying and arranging possible alternatives, the contemporary paint shop manager can effect significant improvements in quality, cost, and productivity, while delivering a product the market demands.

The customer is seldom aware of this part of the iceberg residing below the surface.

Fundamental elements

Figure 2-1 identifies the areas of major considerations in development of a finishing operation, highlighting the physical elements in the context of the human requirements.

As can readily be seen, there are a great number of areas where alternatives are available as one builds the process. Careful study of the chart will help guide the paint manager in building the optimum level of flexibility into any finishing process. Few operating personnel and fewer management staff are aware of the myriad decisions that have been made long before a product finds its way into the finishing area. (A detailed discussion of many of the process alternatives appears in Chapter 10, "Planning Facilities and Painting Strategies.")

A lot of roads lead to the town square. In finishing, several alternatives are available to perform the same work. It is important to understand what these variables do for you.

Implementation methods in most finishing operations should be *evolutionary* in nature rather than *revolutionary*. We must have in place systems that permit management to implement step-by-step improvements periodically that do not interrupt production flow. This type of smooth transition usually goes unnoticed both within the company and by competitors.

For example, when you install a paint booth, make it a bit larger than needed. Begin production with basic hand-held air atomized or airless spray guns. The next level of evolution would be to perhaps move to electrostatic versions of those guns or to HVLP types. The next step would be to consider moving to some form of automatic (robotic) application.

This is a step-by-step improvement that can be implemented over several years using much of the original investment. In this scenario, it probably wasn't necessary to change the conveyors or paint handling equipment along with the basic paint booth because by making the booth a bit larger in the beginning, it is sufficient to accommodate the additional size needed for the eventual automatic equipment. Even if you never get to the automatic stage, incremental change pays because it is not uncommon to see finishing requirements change as products change in size or shape. Sizing booths and openings a bit larger or leaving room for an extra stage or two in the pretreatment area is good evolutionary management.

The alternative, by contrast, is an expensive remake of a finishing line or a total and sudden overhaul made necessary by lack of foresight.

This is not to say there won't ever be a need for major reworks in the paint shop. Every 10 to 15 years a revolutionary change will likely be in order, driven

Basic Design Considerations for a Finishing Line

I. Reason for coating
  A. Appearance
  B. Corrosion protection
  C. Handling-surface protection
  D. Insulation-conductivity
  E. Machining
II. Type of substrate
  A. Metal
    1. Steel
    2. Aluminum
    3. Zinc-coated
    4. Brass
  B. Plastic
    1. SMC
    2. RIM
    3. RTM
III. Type of coating
  A. Liquid-spray
    1. Solvent
    2. Waterborne
  B. Powder
IV. Type of Pretreatment
  A. Hand wash
  B. Degreasing
  C. Chemical conversion coating
V. Type of primer
  A. None
  B. Dip
  C. Spray
  D. E-coat
  E. Self-etching
  F. Autophoretic
VI. Application method
  A. Manual or automated
    1. Air spray
    2. Airless spray
    3. Air-assisted airless
    4. Electrostatic
    5. Flow coat
    6. Dip
    7. Fluidized bed
    8. Rotary atomized discs or bells
    9. HVLP
VII. Curing
  A. Air-dry
  B. Direct or indirect convection
  C. Infrared radiation
  D. Combination radiation/convection
    1. Turbulators
  E. Electron-beam
  F. Ultraviolet
  G. Vapor condensation
VIII. Repair-preparation
  A. Sanding
  B. Masking
  C. Blow and tack
  D. Stripping
IX. Production rate
  A. Number of parts
  B. Number of pieces
    1. Assembled
    2. Unassembled
  C. Hours per year operation
X. System flexibility
  A. Future production
  B. Product changes
  C. Sequencing
  D. Scheduling
  E. Number of colors
  F. Multitone capability
  G. Adaptability to different generic paints
XI. Environmental considerations
  A. Air quality
  B. Water quality
  C. Plural components
  D. Clean room
  E. Eating areas
  F. Locker and shower needs
  G. Respirators/uniforms
  H. Grounding/safety needs
XII. Paint delivery systems
  A. Pressure pots
  B. Circulating systems
    1. Tanks
    2. Tote tanks
  C. Material transportation
XIII. Energy sources/availability
  A. Air
  B. Water
  C. Gas/propane
  D. Fuel oil
  E. Steam

Figure 2-1. Careful planning in the design of a paint operations area smooths implementation.

by new technology, regulations, and competitiveness. These, of course, disturb product and people. If flexibility has not been built into existing systems, paint managers will find it difficult to implement evolutionary change. This allows market erosion to creep in, and if a major change is implemented to try to catch up, there will be few secrets. Competitors will be alerted, and will react to better your goals, leaving you behind again, with your efforts negated and your budget gone.

Five fundamental drivers spur changes in our work:
1. A desire to obtain the same quality for lower costs;
2. A desire to obtain better quality for the same costs;
3. A desire to obtain better quality and better costs at the same time;
4. Changes mandated by regulations or new products;
5. New procedures to either re-establish control or meet competitive needs.

These contribute to the total performance of any operation. The first three have always been the easiest to meet and document. They make up the major efforts by most operations that succeed.

The fourth has been a more recent need and will impact finishing operations heavily in the future. Management tends to view regulatory changes at least preliminarily as cost penalties. They often feel "put upon" by intrusive regulatory bodies and resent having to make changes. Experience shows us, however, that their fears are unfounded. These regulatory pressures force companies to look at every step of their processes and, in so doing, most have found offsetting cost improvements. Often, even total cost reductions have been created.

The fifth item is becoming equally important in any long-term strategy for finishing. It takes keen insight to understand this driver to recognize when it is required. And once again, it is different for each operation.

The hazards of status quo

Where plants are governed by long-established work rules, there tends to be a slow erosion of competitive position to someone else. By maintaining existing systems for long periods of time, firms find the rights and controls required to manage are diluted or bargained away. The best most can hope for is the status quo. In reality, something is lost, and sometimes this something can be substantial.

Here is where leadership *must* exert itself and find the ways to modify and lead companies into processes that keep the company competitive. Few would argue that the average firm does not want to improve. To waffle is to fail; a weakened company cannot continue to provide the corporate security needed or quality products for long-term survival. Its fate is generally determined by competitive predators.

Corporate behavior

A sound operating philosophy must include those elements that make a company a good citizen in the community and that combine to promote the long-term interests of the entire organization. Politics or self promotion of any part at the expense of another is certain to destroy a company, its products, and its work force's future.

A disturbing pattern of the past decade has been the increasing absence of participation by management at all levels in community activities. For various reasons, many managers now seek to live in places away from the city in which they work.

There is also increasing career movement by managers from one company to another. The former method of working up the career ladder in one company over a long period of time has given way to the absentee or mobile manager who has less time available for community leadership.

It has been said that "power is corrupting and almighty power is almighty corrupting." This has been proven in all too many cases at all levels. Any philosophy to run a winning organization must include a policy of integrity and a method of communicating policy by deeds and precept. This is one of the pillars of business success, and certain essentials are requisite to that success.

People must be treated in an open, up-front manner. In dealing with my suppliers, I have always told them that I cannot guarantee them business, only an opportunity for business. Invariably that was enough for the good ones, just a flat playing field. They knew where they stood at all times. They were given a beginning and an end time frame. They were given information as to what happened during that period and the results of any findings developed. They were told why they did or did not win the business. Sure, some were disappointed, but we kept our relationships open for an entire career in most cases. The losers knew they could participate at the next round of activity if they wished.

Included within any relationships with people must be the capability to provide for examination of potential grievances — a sort of court of appeals. This function should be provided to all who wish to come forward if they feel some injustice has been performed somewhere. Although in certain instances, upper management might feel that the paint manager should not become so involved, results have proven the validity of the practice. Getting involved is a manifestation of the manager's conscientiousness. If it involves finishing, then it requires resolving, or it will potentially undermine the credibility and performance of the operation and contribution.

When somebody comes to you about solving any problem, it is one of the best compliments that can be afforded. It is the acknowledgment that you have some talent and capability beyond what they possess. So, problem-solving desire and capability must also be pillars of paint shop management philosophy.

The manufacturing mix

Any business is made up of what can be called the "M & M" recipe. Each part consists of an entity beginning with the letter M: methods, materials, machinery, money, management, marketing, and manpower. Although some might take exception to the term *man*power, style is a part of any philosophy, and the alliterative value is lost if the more modern terms of labor, people, or other descriptions of the human element are used. It is the spirit of communication that is important, not the letter of it, yet there are those who continually put "spin" on words, seemingly only to confuse. For example, if you have ever been rationalized, it probably means you are out of a job. If a capacity study has been made, your plant is probably being closed. We need to make a better effort to make communication a part of our philosophy.

The critical component of finishing philosophy after selection of materials, machinery, manufacturing methods, etc. is the manpower or people portion. This is the very heart of a winning organization. Talented and well motivated people can perform miracles.

I learned a long time ago that if you want good sausage out of the grinder, you have to control the pork going in. You can add sage or pepper after the fact, but no magic happens in that old grinder. It just chews up what you put in. You have to have control over the main ingredients, how you are going to introduce them, and how you are going to use them. In the paint shop, quality people provide this, in the framework of an effective organizational structure.

Management's job is to select and train people if control is to be obtained. It must give the direction and provide the goals under which it is to function. A previous success in business is no guarantee of similar success outside of business. Thus, one cannot manage a paint shop like an assembly line or rely on the same skills. Needs must be tailored to finishing.

As a paint manager, you must be prepared to recognize that management is a combination of science and practice. You must be prepared to manage business and manage other managers and workers along with their actual work. Importantly, a very humanistic understanding and consideration toward people must prevail. The "bull of the woods" style is pretty much dead.

Corporate commitment

During my four decades in the business, I have found that American management maintains certain postures, but too often fails to back those postures with long-term support. We have become a nation of short-term performance without long-term vision.

A common posture of today's managers is to stress the importance of quality. We must do everything possible to prevent defects and deliver quality in the paint shop. But when someone sees something not properly prepared or scheduled wrong and stops the conveyor to correct it, along comes management to tear into the person for slowing "productivity."

This same scenario can be carried over into dozens of examples ranging from a poor air cap, bad robot program, leaking pumps, bad filters, improper chemicals, and so on. What's needed in the modern paint shop are managers who back up in action what they profess to the world. Otherwise, no paint shop will be able to react and produce high performance.

Much too frequently, communication rides a one-way street from upper management. People are seldom challenged to think, or do, or exercise their creativity. This suppresses ideas and dreams. The need for people to dream is a vital part of any business philosophy; without it, no creativity exists.

Once leadership has established the direction and tone of an enterprise, management's next function is to make a productive enterprise out of its human and material resources. If it is done well, it will produce something completely different from the sum of its original parts and be greater in value than this sum. This cannot come from inanimate resources such as money, material, or machinery. It comes from that great human resource.

The following chapters of this book detail proven practices and philosophies essential to successful operation of both large and small finishing operations. In them, one common theme recurs: people and how they are motivated make the difference between winning and losing in finishing operations. How well these philosophies are employed and communicated determines their effectiveness on the paint-shop floor.

It is important to take a close look at your company and determine what role the finishing operations can play in total company performance. Few products can be produced and marketed today without having undergone some type of finishing process.

It is necessary to look both above and below the obvious visual levels and understand the flexibility that can be provided when designing a finishing operation. There are philosophies that lead us to do the proper things, not only for our product, but for the people and society associated with our efforts.

Evolutionary or ongoing improvements must be a part of any philosophy of management that aspires to achieve long-term viability for its enterprise.

Organizational structure must provide the ability to control the primary ingredients utilized through a well chosen and highly trained human resource.

Leadership must provide the tone for goals and direction so management can make a productive enterprise out of human and material resources. If this is created and held responsive to basic beliefs and philosophies, great added value will be achieved. Integrity must be both provided and demanded in all dealings and communications.

CHAPTER 3

Teach Me How to Fish

No one succeeds in the world like the tradesman who has invested the time and effort to learn his craft from the ground up.

There is a story that says bringing fish to a starving person is appreciated, but if you would teach the person how to fish, he or she could provide his or her own food and be eternally grateful for the deed of kindness shown.

Applying the parable in the business context mandates that management control activities, costs, and product quality. To do this, it is necessary that people understand what is expected of them, be trained to do what is expected, and have the necessary tools and support to function. This is what is known as "teach me how to fish."

The currency of success

If there is one attribute we can never acquire enough of in our lives it is knowledge. Knowledge is a dynamic entity. It builds layer upon layer to increase our understanding. At times, certain layers must be discarded and replaced with updated knowledge. Probably 85% of everything I have ever learned about paint materials has been obsoleted by advancements made over the years. It is essential to personal and professional growth to continually acquire and use knowledge.

Knowledge is the currency of success. One could give up everything he or she owns in material things, if necessary, to obtain knowledge and still emerge a winner. With knowledge gained, material wealth can be regained. Knowledge is the grist of the success mill, that which engenders the confidence to build quality of life.

A highly personal possession, and one that is both portable and marketable, knowledge is a uniquely human-owned resource. It does not go into files or on a shelf. It is the product of formal learning combined with actual skills gained by experience.

Because it is marketable, those who have it are actively sought after. The key is to create a workplace environment in which workers want to stay with a company after they have been taught how to "fish." It makes little economic sense for a company to serve as the training grounds for other firms, who then reap the benefits.

The astute manager will work to ensure that all pieces of a good working environment are in place, that communication flows lucidly bottom up and top down, and that knowledge applied results in performance recognized. Just one missing piece will degrade the performance of the whole.

Knowledge at work

No one succeeds in greater measure than the person who has taken the time and effort to learn something from the ground up. No one succeeds like that person who is also able to physically apply what he or she has learned. The formula for success is further embellished by the addition of confidence, patience, and the ability to "speak the language" of the trade. I am reminded of a time I was called to a company to tackle a problem of poor paint adhesion compounded by practically no corrosion resistance.

Most such problems usually are associated with poor cleaning and pretreatment. However, I started my investigation at the beginning of the entire manufacturing process, examining every facet of metals, lubricants, and handling. By the time I got to the pretreatment system, it seemed every client of this company had arrived wanting to know when they could expect product to begin shipping.

The pretreatment log indicated clearly to me what the problem was, yet no one had grasped what it meant. In fact, the log entries were being considered a positive indication instead of the source of the problem: the rinse stage following the alkaline cleaner stage was reading 0.0 total points of alkalinity, which on the surface at least would indicate that the rinse was perfect. Quite the opposite. The only way to get this condition normally would be to test for alkalinity and find none. In other words, the stage was really acid in pH. This indicated a leak from the phosphate stage back into the rinse, creating soluble salts and other problems of coating leading to the failures.

A check of the system revealed the source of leaks and contamination. Corrections were made and the product came immediately back to the performance levels required. In fact, with some other refinements, the corrosion levels were increased to double the original levels and other savings were effected through better controls coupled with better loading patterns. The savvy of the shop floor prevailed to solve the problem.

Time and again this truth has been borne out. Boardroom-style analyses do not soar on the plant floor.

In the fish bowl

People constantly watch the managers. They will quickly pick up any inconsistency in actions or positions taken and reflect those in their own behaviors. If the manager is having a bad day, it will manifest the same negative effect all around. If you as manager look like you are not serious about what is happening, those around you will view things the same way. Conversely, presenting an image of confidence, warmth, and performance will reflect positive effects. You are a mirror that reflects all around you, and the image projected will be the image reflected.

That image is important in dealing with the work force, both individually and as a group. Different types of people are usually found in a mature manufacturing facility. The ability to discern their similarities and differences distinguishes the mundane manager from the effective leader.

Most people in the floor operations of a paint shop or in the other manufacturing disciplines tend to be either "tribalistic" or "traditional" in nature. As I see it, paint people tend to be more tribalistic. (I use this term in its most benevolent sense, that of a close-knit group showing similar traits and interests.) They gravitate to each other for a variety of reasons, drawn by the common bond of painting. It has been said that paint people like to smell the fumes just as machinists like to see the chips fly.

History teaches that tribes expect to have a leader to follow, a leader who has earned the position through various forms of trial. It is much more difficult to have a style of delegating activities in this type of group. It is expected that the chief will lead by example and deed. Applying the metaphor to the paint shop floor, first-line management and all those in authority must physically be able to perform all the required process functions if that person is to fully expect acceptance by the group. Being all theory and no example will result in little buy-in or improved performance.

Buy or build

Based on what I have learned during my career, a company should spend about 1-1/2% of gross revenues on employee training. Unfortunately, very few even approach this level, owing largely to two commonly held beliefs. First, that training is for management only, not floor people. This practice is based on the feeling that a company has little control over how long a worker may stay on a certain job, so the money would be potentially wasted. These firms do spend money for management training in various ways.

The other belief is that companies should scour the market and hire the expertise. They do not want to "grow their own" and are willing to pay higher salaries for what they consider trained and performing people.

Examples to argue each practice are readily available. It is my feeling that some mixture of the two probably is best. Without internal promotion prospects, good people are going to leave any firm, arguing against the practice of hiring from the outside. Conversely, it can be argued that new blood and fresh ideas coming into an organization are good, too.

Companies more and more are thinning out the middle management ranks to reduce costs. This is requiring floor people to take on some elements of management in addition to their regular tasks. The onus of management here is to make available more information about the business and empower people tasked with the new responsibilities. This means teaching other skills on the floor. It means bestowing more trust on people to obtain control rather than be controlled. This can shake up the tribe or it can make it grow stronger.

In the past decade, there has been a growing attitude that management should come from the college ranks, yet few universities offer a degree in paint technology: graduates are geared more toward manufacturing paint than using it. Most of these young people do not possess the experience and hands-on ability to do something around finishing systems. Many lack maturity, have limited communications skills, and haven't earned respect from floor people. They are neither leaders nor managers, yet they are thrust in those positions unprepared. They are fish out of water before hardly beginning a career.

It requires a lengthy time for someone to become reasonably proficient in the finishing area. It takes extensive hands-on experience to understand racking, metals and nonmetal substrates, various forms of pretreatment, materials, application techniques, curing, health and safety, process engineering, budgets, waste prevention and treatment, and a lot of aspects both ahead of and on through the finishing area.

Paint shop managers must be teacher to the teacher. Only then can the teacher possibly teach the student. However, there are those who firmly believe all learning should take place from actual experience.

The problem with that theory is that in nearly every case, the students get the test before they get the text. It doesn't make sense to let our bread-and-butter operations become the training grounds for our learners. It is tantamount to hiring a fishing guide who is learning to fish right along with us.

Good managers use data from operations to formulate justification to their management that money spent on training yields a positive return. When they do this, they are exhibiting leadership. They are influencing the makeup of the work force in a positive way and changing the actions of their organization. This is another form of teaching people how to fish. This is teaching managers.

With the new combined pressures of world-class quality, environmental mandates, and global competitiveness, we will see all kinds of opportunities and challenges in our finishing operations. This translates into a lot of effort to acquire knowledge and capability ourselves and then to share them with those around us. Literally hundreds of paint shops operate today exactly like they did in the 1970s. Too little effort has been made to introduce new materials and application equipment to improve performance and impact on the environment. Management's position in most cases has been to avoid new expenditures and simply "milk" the old ways, wondering why profits shrink and jobs disappear. Such complacency is ill-founded. Many new technologies are available requiring only minimal capital outlays, with training provided by suppliers. Suppliers, however, become increasingly reluctant to develop more productive equipment/training packages in the absence of the market drive to support them.

I am reminded of a company which had gone to a waterborne coating because of environmental considerations. At the same time, the company eliminated electrostatic guns because of the costs associated with isolating and grounding the material lines. The net result was twofold.

First, the company's paint bill soared because the waterborne paint was low in solids and required a great deal more paint to reach the needed film builds. Second, transfer efficiency fell to very low levels compared to those of the former electrostatic application. This combination greatly increased the total paint bill and actually saw as many VOCs emitted as with the former solvent-based finishes.

The company then introduced HVLP application and saw dramatic gains in both paint savings and VOC reduction.

Why did the supplier not resolve this problem or, perhaps more importantly, why did he ever let it happen in the first place?

My view is that this was a classic example of supplier and company management holding the status quo until some major problem forced a change. Then the change was not performed well and the subsequent penalties were considered part of the price for meeting environmental regulations. Leadership and management were found lacking in both the supplier and manufacturer.

The bottom line is that "low-visibility management" and lack of leadership cannot survive in a globally competitive environment. A point will come very soon when environmental laws will force people to either get back into or get out of finishing. Instead of evolutionary changes, there will be the chaos of playing catch-up or the specter of going out of business. Either way, there will be a great need for leadership and vision to prepare the work force of the new century. A big part of getting ready is spreading the word.

Knowledge comes from both formal and experiential education and is only useful when shared. Its dissemination should be to as wide a circle of associates as possible in timely fashion because it is also dynamic, often made obsolete by new technology.

World-class quality demands, environmental compliance, and increasing competition mandate new and expanded roles for our human resources. Continuing skills training and knowledge exchange will determine the success or failure of companies in the next century. Doing the right things at the right time in the right amount comes from leadership and management dedicated to teaching and nurturing people.

The best gallon of paint or most potent pound of chemical is the one you never have to use to get the job done correctly.

CHAPTER 4

Putting Together an Organization

We trained hard, but it seemed that every time we were beginning to form into teams, we would be reorganized. I was to learn later in life that we tend to meet any new situation by reorganizing, and what a wonderful method it can be for creating the illusion of progress while producing confusion, inefficiency and demoralization.

*Petronius Arbiter
210 BC*

It is safe to assume that the chapter lead-in quotation, when written, was not intended for contemporary manufacturing or finishing; it is however, fitting commentary on many of today's businesses.

The musical chairs syndrome

From somewhere came the popular belief that reorganizing everything is the way out of trouble and the path to the pot of gold. If something's broke, reorganize. Then reorganize again. This is short-term thinking with long-term consequences. If I had at my disposal just a portion of the "reorganization" losses I've seen in my lifetime, I could have upgraded every paint shop I've worked with to maximum efficiency, with plenty of funds left over to train people.

During one point of my career, I reported to four different vice presidents in a seven-month period. They were all good people basically but each had a greatly differing set of priorities and plans. We created more plans during this period than I care to describe. And before any could be implemented, another reorganization would take place. Fortunately, for paint operations, I was at least one common denominator as a career professional, and this small bit of stability proved to be enough to prevent damage in other non-paint areas.

Many jobs in industrial organizations are pretty well defined and programmed by rules about what is to be done, in what order, and how quickly. This is generally found in paint shops and their hourly and piece workers are highly programmed. Another throwback to *Taylorism* of years ago.

Strangely enough, the work shown at the top of most organizational charts is just the opposite. It is not highly programmed. These are "think" jobs and much harder to define or describe exactly. Middle management tends to have some of both situations, with the "think" portion limited to some specific rules to follow with varying amounts of judgment and autonomy.

Automated, new technology for dispensing information has been changing many areas of both forms of jobs. Scheduling and functions like timekeeping are increasingly reducing the need for human judgment and experience. Areas such as calculating yield decisions for product mixes, warehousing, budgets, and others are now handled by programs rather than people.

Eventually, however, the use of new technology brings up old human relations problems. On the one hand, advocates of new techniques feel that a depersonalized, highly programmed, and machinelike organization produces a better, less political working environment to manage. On the other, morale and pressures to conform to changing requirements can impact productivity.

The changing face of management

My experience has shown that the individual middle manager has personal aspirations and needs to satisfy, but now he or she usually meets these needs off the job, just like many of the rank-and-file have done. When this occurs, the individual derives less satisfaction from within the organization and what naturally follows is less dedication to the job and the company. The focus moves on to other noncareer interests. This approach only succeeds in making middle management less dedicated to introducing technology.

If pathways to innovation are blocked, people become impersonal. One should use the time freed up from use of programs and computers toward more planning and problem solving. Innovators then must be supported by two additional types of people: those who will take ideas, approve or veto them, and obtain the necessary resources to put the approved ideas into operation; and those capable of physically doing the work to implement ideas after approval.

I see companies doing some of these activities quite often, but seldom do these same companies provide the approval and implementation pieces required to turn the ideas into real payoffs. Eventually, the "thinkers" give up or move on to other companies who just might be more receptive to their ideas.

One of the unfortunate things that often happens in management's quest for improvement is that it ends up being destructive to the organization. A manager will sell his or her people on becoming partners in improving performance. In team fashion they together achieve their goals in the paint shop. But the typical reward in the past has been layoff notices for some of the people because they were no longer required. They lost their jobs for having participated, and the others left on the job then began to look elsewhere for their needs and focused loyalty on other options. It is a Pyrrhic victory to achieve efficiency but lose the organization. You can't use people for the basis of your improved record and then discard them.

This mobility of managers will impact new compensation plans and performance appraisals. The old Tayloristic ways of payment from highly structured and measured work performance will also change where the work force is concerned.

Management, to be effective, must take a long look at itself from an internal and psychological point of view. Information technology will certainly change many established practices, and we will have to rethink attitudes and values. The work of the individual and mobility of management will necessitate this. It does not take a long suit of knowledge to lay someone off. But it does take wisdom, skill, and planning to guide a business such that it avoids losing its human resources.

Organizing the organization

Certainly, organization is necessary for order and operation. It is necessary to evaluate performance, meet the needs of the marketplace, and let the world at large know who you are and what you do. But what are the ingredients of a good organization for a finishing area, and where does it fit in the overall organizational structure?

Structure of the organization should contain certain elements and their respective talents. First, my experience has taught that finishing should be a totally separate and identifiable area. It should not be under a part of metallurgical or support areas like quality or maintenance. When such is the case, it almost always becomes a stepchild to the other functions.

Where does finishing fit?

Finishing should be one of the mainline disciplines required to perform the total manufacturing operation. We have previously identified the many areas paint and finishing can influence in design and manufacturing. Now it is necessary to integrate this function into the framework of the many disciplines comprising the overall manufacturing process.

Whether in a large operation or a small one, the more you can make finishing a self-supporting entity, the better. Paint and finishing is not like other disciplines. It is so much better if people can have a compact and highly directed focus on the needs of finishing.

There must, of course, be a manager. This person may be called by any one of a number of titles — general foreman, superintendent, area manager, and so on — but the important thing is, that one person has the overall responsibility for finishing. (See Figure 4-1.)

In large paint operations, there will usually be an operational manager, charged with the actual running of the area, and some form of technical supervision on equal standing with the production manager. In this structure, both must relate to each other's needs and neither has veto control if they have conflicting views. Veto is left to the general manager.

In a smaller operation, one person may serve both roles, but a potential danger therein is that sometimes the technical side suffers, causing on-going performance variations.

I say this because of my observations over the years of finishing managers who were told they were responsible for their operations, yet really not in control of the various aspects that influenced their performance. Products were designed by product engineering; the systems were supplied by manufacturing engineering; materials were acquired by a purchasing department; process routings were made by a planning department; time study by industrial engineering; and the quality levels by quality control. It seemed everyone outside paint had more influence but little final responsibility.

People in all these disciplines tend to view finishing in a different way and have their own interests at heart. It becomes necessary, then, to form a philosophy that insists that paint management actively input the technical data upon which the supporting service areas act. Without this, it just gets lost too often in the profusion of other activities, whose management is not technically oriented toward a finishing area.

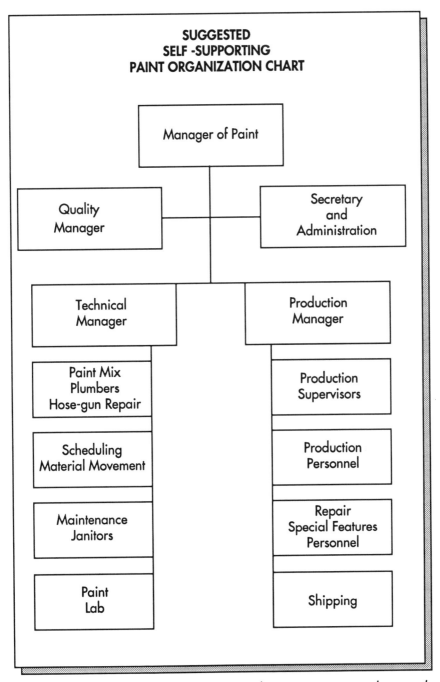

Figure 4-1. In paint operations, the more self-supporting the management structure, the greater the focus on the functions of finishing.

The finishing area should have in place its own material handling, maintenance, safety, and housekeeping functions under the purview of the technical supervisor. Although not direct production people, but as the focal point for process and technical control, the technical supervisor and staff support the production manager in every way.

On balance, the needs and responsibilities of the technical supervisor and the production manager should even out. Each must contribute to the whole. The buck stops and performance improves. We have now made the system responsible for its decisions and the kind of sausage that comes out of that grinder.

As we stated earlier, finishing is like an iceberg. It demands attention to basic hour-to-hour details that must be committed to by management. You have helped answer questions, like How good is good? How much does this method cost versus alternative methods? How much do we maintain? How did we schedule and present product? Did we protect it from damage? Did we supply burden materials, repair, remove trash, empty containers? These and a dozen other things are part of the unseen bulk of the iceberg.

The management/staff mix

The managerial portion of the organization can now be completed with the addition of floor supervisors who are experienced "doers." These are people who have usually come from floor experience to management. They represent stability in the organization and are happy in their expertise. They also represent the best choice for the next generation department head or mid-level manager.

Next, there needs to be a means for selecting and qualifying people to perform the everyday functions of actual finishing. This can be either a very structured approach mandated by union contractual language or the opposite, where hiring and placement are not limited in any manner. The success or failure of either can be explained with an understanding of some of the basic personalities found in people.

The influence of technology

Technology is beginning to change the look of management charts. Many middle management jobs are changing, like the transition from old-time craftsman to today's piece worker. (See Figure 4-2.) As automation occurs, the need for management to be "doers" of the old techniques diminishes, but that

THE CHANGING APPROACH TO ORGANIZING WORK

What Management Assumes About Workers

OLD WAY Worker wants nothing from the job except pay, avoids responsibility and must be controlled and coerced.

NEW WAY Worker desires challenging job and will seek responsibility and autonomy if management permits.

How the Job is Designed

OLD WAY Work is fragmented and deskilled. Worker is confined to narrow job. Doing and thinking are separated.

NEW WAY Work is multiskilled and performed by teamwork where possible. Worker can upgrade whole system. Doing and thinking are combined.

Management's Organization and Style

OLD WAY Top-down military command with worker at bottom of many supervisory layers, worker is expected to obey orders and has no power.

NEW WAY Relatively flat structure with few layers: worker makes suggestions and has power to implement changes.

Job Training and Security

OLD WAY Worker is regarded as a replaceable part and is given little initial training or retraining for new jobs. Layoffs are routine when business declines.

NEW WAY Worker is considered a valuable resource and is constantly retrained in new skills. Layoffs are avoided if possible in a downturn.

How Wages are Determined

OLD WAY Pay is geared to the job, not the person, and is determined by evaluation and job classification systems.

NEW WAY Pay is linked to skills acquired. Group incentive and profit-sharing plans are used to enhance commitment.

Labor Relations

OLD WAY Labor and management interests are considered incompatible. Conflict arises on the shop floor and in bargaining.

NEW WAY Mutual interests are emphasized. Management shares information about the business. Labor shares responsibility for making it succeed.

Figure 4-2. Technology and workplace flexibility mandate changes in how we manage work and prepare workers for it.

for knowledge of programming and other sophisticated tools increases. Some managers will be pushed upward to plan and program, while others will be reduced to routine policing, or be eliminated completely. Those who remain will have reduced responsibilities as "thinkers." Like the workers who receive less satisfaction on the job, lower management may find itself seeking accomplishments outside the work environment. The functions of management do not require great enthusiasm; they must be changed to something more fulfilling.

Another possibility is that we may wind up with a new level of elite technical management personnel whose very survival depends on their ability to stay abreast of technology. Those content with status quo could well be faced with the prospect that they themselves will become obsolete within their own career's time span. On one hand, the top will have to think more to be competitive, but on the other, they will increase the rate of burnout and turnover of the innovators.

The entire idea of turnover, compensation, and pensions will begin to change as "brain picking" becomes more pronounced. This is why it is my belief that the healthiest and best performing organizations are those based upon some combination of traditional grooming internally as the primary supplier of knowledge and some new infusion from outside. My experience has shown this to be on the order of 80% to 85% internally and 15% to 20% externally.

Whatever the people need, either management or floor, I always look for people exhibiting the sixth sense: "common sense." Without this, the other senses mean little. But combined with personal integrity, it is a winning formula.

Too many times cleverness is substituted for loyalty and talent. When put to test, an ounce of integrity is worth more in management than a ton of cleverness. Not blind loyalty, but the old-fashioned variety where one is willing to risk reputation or incur a little wrath to do what is proper.

A brief but powerful moment occurs in life that transforms people into performers. It is that point of realization when they have just made a decision and accepted the responsibility for that action. When we act in such a manner, a simple and important message of commitment is imparted. I believe in looking internally first for this human attribute.

The experience of one of my clients is exemplary of such a transformation. This client had submitted to an organizational change along the lines shown in Figure 4-1 and had established the position of corporate paint manager. The new corporate strategy also called for using electrostatic application. One of the plants which up to this time had always set its own agendas balked at the use of such equipment. A well-formulated and firm response was generated to answer this first challenge to the approach. When given in a confident and professional manner, the matter was no longer an issue.

It is doubtful such a response would have happened before the transformation of the new manager had turned him into a high performer.

But where do you get these kinds of people in the shop environment, especially when many operations feel strangled by contracts and seniority agreements and workers are constantly transferring from job to job? Although these circumstances are not ideal, there are ways to do it. It is a bit slower, but you can attract and hold good people in the finishing areas in either mode. It takes a good understanding of the types of personalities and attitudes present in a potential work force.

The social mix

Popular today is a movement which states everybody is unique and distinct. This may be so outside the plant or office in private lives, but it would be unmanageable in business if everybody did their own thing when, where, and how they wished to do it. That's why organization is essential to provide various levels of control. Distinct behavioral patterns exist at every situation where people are involved. I have found this to prevail at the floor, management, executive, and even customer levels. Let's look at what experience teaches.

Approximately 35% of all people are metaphorically tribalistic in value systems. They believe in their "chief," and when the chief says it is so, it is so. It is next to impossible to get the followers up to higher levels of productivity through modern methods. Job enrichment and offers of promotion will not do it. You must get a better chief, or leader, with higher goals, more strengths, demonstrated performance and courage to set examples for these people. They tend to be suspicious of formal presentations and react best to graphics and other forms of communication. Cartoon safety and quality posters decorating plant walls today attest to this axiom.

These workers are usually quite content with routine jobs on presses and other heavy work situations. In the paint shop, you can find them in manual spray areas. Management has to be able to outpaint the employee to earn their respect and to move them toward goals and new methods.

Another 35% or so of people are conformist in their nature. They tend to respect tradition in a company and community. They are loyal, believe in formal policies, the printed word, and require less personal leadership by management. You communicate to these people in a conventional manner or with formal announcements on bulletin boards or letters mailed to their homes. It is from this group that your best frontline floor management can be recruited. They believe in this type of progress in the business world, from the bottom up.

Sociocentric people represent another 10% of the population. You will generally find them more in clerical and scheduling areas. They also enjoy running computers and other electronic replacements for past clerical work. They wish more control over their own actions and assignments and want little direct supervision, a friendly and open type work atmosphere, and few rules. They tend to carry causes into their jobs, however, and will often be the most vocal about the company and how it is treating employees or impacting society. While they can perform certain activities well, you perhaps would not want to give them a great deal of formal responsibility and authority. They are reluctant to accept it and will likely never make tough decisions or accept risk.

Such people do fill an important need in an organization. They work well alone, at night, in testing and evaluation, or in any data generating function. They lean toward being loners, and subscribe to the Just-tell-me-what-you-need-done-and-leave-me-alone-I'll-get-it-done school. Their need for this solitude is the force that drives them to perform for you so that they can perpetuate their working situation. They make good workers for astute managers.

The next two categories are those in which management would score fairly high. About 15% of people are manipulative in nature. They hold goals, incentives, promotions, and ways of getting ahead as important values. They try to create policies and rules to guide their own and others' activities. Within this group is a subset that can be both productive or destructive, depending upon how they are directed.

Workers in this group aspire to do things better, which is good; unfortunately, many of their goals are individualistically motivated rather than being directed toward a total team or company concept. They are opportunistic and can be politically dangerous. They are capable of betrayal and sabotage if it promotes them personally. They are masters at reading the pulse of situations where they can turn their own goals close to those of the company and gain for themselves a rapid rise to the top, skyrocket style.

It is good policy to stay clear of these individuals when staffing. Paint management has to be a professionally trained, career oriented, ongoing type of stable organization. Continuity is a must because most paint materials require many months of testing and approval for use. The "overnight skyrockets" can kill long-term efficiency in a paint shop.

In my estimation, two kinds of people populate most business organizations: those who create assets and leave things better, and those who use assets for personal gain and leave things worse. Behind the *creators* of assets you'll find a healthy, thriving company. Behind the *users* of assets, all you usually find is the burned-out skeleton of an organization. If I were to classify paint managers, I'd have to put them in the asset-creating category. Few paint and finishing managers ever rise through the organization to executive levels. They seldom ever really expect to do so. They are paid professionals and not likely to become self-serving users of assets. They instead run their organizations with the goals of the company in mind, contributing to profitability and corporate health.

A last category representing about 5% of all people can be found in all areas of life. They are the ego-driven group who only respect power. They are tough, aggressive, and never satisfied. They like nothing, and are impervious to anything said to or about them. Such people cause 95% of the grievances and problems in a work area, and generally tear down rather than build up. Often they make or break management and organizations. How you as manager handle these people and the difficult situations they create will be carefully noted by your entire staff. They will react based on whether you temper the situation or submit to it.

My grandfather was a blacksmith by profession, a lay preacher by calling, and a southern Ohio philosopher by avocation. He believed people were either like butter or iron. "The same fire that melts butter will temper steel," he would tell me when I was a child. Though I knew this even at that early age, it would be years before the full meaning would become apparent to me. The old man felt that people had to go through the heat and pressures of life and get pounded on a bit as a part of the rite of learning. If you were made of the right stuff, you would grow strong and yield new and better properties from the experience. If you were made of butter, you would just melt.

There is no universal formula for developing good management technique. It is a learn-by-doing proposition. I recall from my experience a case in which I had the dubious privilege of having in my paint lab a paint technician who was long on seniority but short on integrity. He had some talent and capability, but also had difficulty keeping focused.

Over time, this person came up with more sicknesses and other ways of being away from work than I had ever before encountered. It seemed he was always involved in some scheme or another. Fortunately, I was able to intercept him in most of his schemes and head them off before they did real harm.

Although this behavior resulted in a waste of his talent and my time, some positive results were produced: by having to deal with this worker's waywardness, I was able to hone my skills as a day-to-day manager. I gained insights which I might never had developed without the challenge of his manipulative mind.

On another occasion in a different paint shop, there was on staff a person who was seemingly slow on the uptake and, consequently, was the brunt of a lot of teasing and ribbing.

When one day the shop manager was to be away from the plant, the paint staff thought it would be great fun to put this person in charge. I didn't agree. In fact, I felt it was time to perhaps turn the tables a bit.

During my brief exposure to the "victim," I had found him to have a deep yearning to please and perform. So I took him aside and instructed him in some basics on how to assign people, do time cards, and schedule work flow and materials. To the astonishment of his colleagues, he performed well; so well, in fact, that by day's end, they were helping him. This person, when given the chance, exhibited new capabilities that could well have gone unnoticed forever.

Nurturing the human resource is a critical part of management responsibility. Taking the risk, in this case of making the person a manager-for-a-day, not only uncovered latent skills in the individual, but fostered team inspiration as well. Leaders have to take risks at times.

It is from the heat of these experiences that organizations and people are either made to have new and better properties or broken down and melted away.

The new paint manager

Management has to be different things to different groups and different people. To be all things to all people is to fail. There aren't enough hours in the day to hone all the distinct attributes of people; management technique must be developed around these distinctions.

Obviously, it is both normal and necessary to have some form of organizational structure to provide order and method to a finishing operation. It is necessary to evaluate performance, provide value added to the company's products, and stand up and be heard about who you are and what you do.

Structures will look different from operation to operation, but the most effective contain similar characteristics and capabilities. They are self-supporting in nature and have their own identity within the total organization. The self-supporting feature provides a much higher level of control and accountability.

Emerging technologies and the widening scope of markets are changing management philosophies and structures. New knowledge must be obtained and employed. Managers need to be flexible to ensure their viability in their organizations and job markets in general. Those who are reluctant to share decision-making authority and use of controls with floor-level operators will find little room for advancement in the twenty-first century manufacturing world.

Recognizing that people have distinct characteristics which influence their behavior in the workplace, managers must continually and routinely hone those skills that identify the differences to better select, train, and build an effective human resource.

CHAPTER 5

Poor Management, Entitlement, or Work Ethic Problems?

A decade of mergers, downsizing, takeovers, and restructurings has broken the loyalty of the work force.

Manufacturing today is in a wave of heavy layoffs of younger service employees and supervisors with the result being that we probably have more experienced people working today in proportion to the total force than ever before. The logical conclusion, then, is that our operations are capable of turning out excellent product in a timely manner cost effectively.

However, that isn't really happening in many cases. A number of other factors enter into play. As automation and other new technological advances come on line, the young get laid off. The older workers, comfortable with their knowledge and experience, often are reluctant to learn the new skills necessary to support new technology. This reluctance, in turn, neutralizes their years of experience. What needs to be done is no longer done the way they learned it. Add to that a growing reluctance by management in mature plants to hiring young workers, and we have the dilemma of disappearing job skills in a work force with high entitlement expectations.

When we move young workers into new plants with no seniority provision, we are faced with a completely different employee than we were 20 to 30 years ago.

The now generation

The new breed of Americans today was raised within the psychology of entitlement. Theirs is the feeling that they are owed a college education and entitled to a meaningful or at least interesting job.

At the workplace, they likely make more money earlier than their fathers, get more vacations, and are even able to retire earlier. When they come up against reality and find themselves glued to what they perceive as a meaningless job, they feel entitled to gripe about it and seek redress. Retirement is a long time away, and the "now" is important.

Two major changes have taken place in the work force in the last 10 years. First, the greater self-interest of all employees has given rise to the "me" syndrome. Second, the increasing level of education and the growing numbers of women and minorities in both labor and management are changing the environment for people and their work place.

The young employee tends more to question the authority of supervisors or even unions as a matter of course. This now generation expects to participate in decision making. The young are challenging the entrenched older leaders of both companies and unions. More and more they challenge bargaining agreements as negotiated by their leaders. They do not accept "no can do!"

The work force is now made up of more than 50% women who express still different concerns about their employment. There are also many two-income families today whose combined income is often greater than the older traditional supervisors with no working spouse. The new worker enjoys the good life. Job enrichment, leisure time, and even maternity or paternity leaves are today strong elements of their compensation demands.

The corporate squeeze

In direct opposition to this new wave of demands is industry's move toward cutting back on entitlements or benefits. We are being told there is a declining work ethic in the United States *vis a vis* the Japanese. We are told most Japanese work six days a week, and that half don't take all the vacations due them. We also read that thousands drop dead at their desks, succumbing to overwork.

American industry has responded by working people more overtime in an attempt to increase profitability. But productivity has suffered. We have created a work force characterized by burnout, alienation, and under appreciation.

This overwork is brought on by the economics of the workplace: it is less costly to provide health and other benefits to fewer employees. But holding down employment only widens the gap between the out-of-work and the overworked. Those not working are missing out on the American dream, and those who are overworked are working harder for the same lifestyle, with less time to enjoy it. In both camps resides a threat to the American work ethic.

I do not believe that the American worker has lost the work ethic. Our problems stem from the way we manage the workplace. There is a deteriorating relationship between the worker and the workplace. A decade of mergers, downsizing, takeovers, and restructurings has broken the loyalty of people. Neither the losers nor the survivors are apt to direct their loyalty toward an organization that deals with people as numbers to be manipulated on the balance sheet.

Profit philosophies and cost shifts

It has been my good fortune to have worked in many places around the world. This has given me first-hand knowledge of the so-called work ethic and the opportunity to compare productivity under different management techniques. If we use our oldest and best asset, American ingenuity, as a measure, the average worker here is as productive as any in the world. The problem in recent years is we have stopped becoming innovative and started chasing short-term profits instead of long-term goals.

This raises questions about why American industry and management have stopped investing in our own nation and future and why we continue to chase cheap labor around the world when there is technology available to retain jobs and manufacturing here. In the early 1950s, it was not uncommon to find labor representing about 65% of the costs and materials about 35%. Today, it is more common to find materials costing up to 90% and labor the smallest part. This is truly borne out in the finishing business. In my lifetime, the costs associated with finishing have done almost a 180-degree turn.

This shift has mandated better labor to run the equipment and preserve materials, not cheaper labor. A well trained person can save a big portion of labor costs through material use improvement and reduced rework in the finishing area.

The U.S. Labor Department states that the average Japanese worker puts in nearly 200 hours more work a year than a U.S. worker. But hours on the job do not necessarily equal productivity. If these are productive hours, it is due to new management techniques and philosophies, most of which were originally developed by Americans.

The Japanese, if nothing else, have been quite astute at recognizing management philosophies that work. These philosophies have led them to be more technologically adaptive than U.S. managers. Where in Japan, buy-in by all concerned was almost automatic, here management feared technology because it depended more on workers and sharing the decision-making process. We did not see the changes unfolding, and it is still easier to ship work elsewhere than take risk, invest, and learn how to apply new management philosophies.

The U.S. is large and endowed with abundant resources. Technology is available to compete anywhere if management and leadership make it happen and are willing to change outmoded chain-of-command perspectives. Too many companies have a "Charlie Tuna" management philosophy. They are more interested in image and providing tuna that has "good taste" than understanding that the market wants tuna that "tastes good."

Old ways, old results

In my travels throughout the world, I have had the opportunity to observe the operations of major producers. Interestingly, the manpower assignments were not close to what would be typical in a U.S. plant. In fact, the workers probably would have struck their plants if a workload comparable to that in the U.S. had been placed on them. Materials as well were not as good as found here, and they cost far more, as did energy. Where they excel is not at the worker level, but in the use and management of technology, material handling, space, supporting services, and attention to management philosophies of first-time quality and reducing waste. Management has recognized the value of efficiency in these areas and trains and leads its people accordingly.

Illustrative of the need to overhaul some of our management practices and policies are the authority levels granted to spend money. When I entered the manufacturing world in the early 1950s, a plant manager was permitted to spend up to $1,000 without going through the approval process. Today, 40 years later and a with a dollar worth a quarter that of 1950 levels, it is still at $1,000 in many companies. That is not much trust given for the results expected.

What is needed today is a realization that the old ways don't work. You get old results out of old ways. We have to understand the system can't support a way of life built upon entitlements. David Halberstam talks about entitlement in his book, *The Next Century*. He says, in part: "It is the inevitable product of three generations of affluence, which in turn created a culture of high expectations, which in turn created a politics premised on high assumptions and high consumption. It turns on the idea that everything is as it used to be. The result is a society oddly oblivious of its new realities."

The marketplace in which we compete does not operate under rules of entitlement. If you don't provide the highest overall value to the customer, a competitor will take over. Ultimately it makes little difference what we think we are entitled to; if we can't pay for it, it doesn't matter.

Management and leadership strategies that shrink companies and target only short-term profits and considerations cannot expect to make anything meaningful, let alone entitlements. In this no-win situation, workers usually end up with 100% of nothing at some point, and companies deteriorate.

The reality is that we are different from the Japanese and other cultures in the world. Our greatest strengths come from the diversity and friendliness of our people. We share not only things in this country with others, we also share the openness of our country to the whole world. This great diversity, channeled correctly, is the fertile breeding ground for creativity. We can draw on the best of any nation here. The U.S. has the most unrestrictive environment, both politically and culturally; this alone allows ideas and creativity to flourish.

In truth, the Japanese and others have studied and copied us on the way to their industrial positions today. We served as their benchmark and still are the greatest economy in the world. What has happened is that management has just turned its focus to short-term profits in just about every phase of business. This redirection has not proven to be the basis for success at any level in the enterprise: in the finishing shop, the division, or the company as a whole.

If I was beneficiary of any success during my career, it was due to thinking just the opposite. Management planned its operation on the basis of creation, evolutionary improvement, and eventual replacement of old technology. It was long-term stability of ideas, knowledge, and integrity that people could understand and depend upon, even at those times when people did not like what they were hearing. Our decisions were based on the truth and in their best long-term interest as an organization. It was not easy at times. Eventually this stability of long-term involvement created a professional relationship and ability to communicate to various entities both inside and outside the organization.

People never looked at me as trying to become the next plant manager, the next vice president, or CEO. It was never my feeling that not being one of these constituted a failure. What's wrong with being a good paint and finishing professional? What's wrong with being the best in your corporation or one of the best in the industry worldwide? Those are the goals I sought and focused upon, not building a resume by moves every few months.

The American challenge is to communicate a willingness to be flexible and adapt to new challenges and approaches to operate our finishing shops. If we are entitled to anything, it is the chance to be the best we can. People are not losing the work ethic, they are losing guidance and leadership to management's short-term performance goals. In company after company people are renewing their quest to being the best in the world when given effective goals, leadership, management, and the will to succeed. Myriad success stories around the country attest to this.

The challenge of change

We are undergoing great change and stress in a changing world. It is becoming more difficult to find security in the work place today for a variety of reasons. This places both the mature and younger person in scenarios not experienced by past generations.

Problems of attitudes abound in both workers and management. Shock and disappointment are visibly expressed when workers come to the realization that most of their preserved entitlements are myths when reality sets in. Their wants and needs are further complicated by the changing psychological makeup of our work force today compared to the past.

While people have been seeking such perks as job enrichment, leisure time, and various types of leave, industry has been downsizing and working more overtime to increase profits for the short term. Generally, this action does not bring about a harmony of interests between the parties. It is not different management, but the same kind of management, on a smaller scale, looking to the same outdated management forms for answers to new problems.

The result is breakdown, the reaction to which is to demean the worker as having lost the work ethic or being unqualified. Some managers fear technology and change because they place more emphasis on talented labor and sharing of responsibilities. In the finishing area, these fears are unfounded because emphasis and costs have moved more to the materials side of the ledger rather than labor. Industry needs good craftspeople.

We tend to spend too much time comparing ourselves and copying other cultures, trying to emulate them when in truth they have really copied us. We have to return to our traditional strengths of ingenuity and creativity. We are capable of competing and winning if we combine technology and creativity with the judicious use of the human resource.

Redirecting good management toward long-term goals will resolve most financial and cultural problems in the future. The American work ethic has not disappeared. It just needs to be resurrected through careful direction, guidance, and leadership.

CHAPTER 6

Staffing the Workplace in the Seniority Game

Workers are attracted to management that has sound values. They gravitate toward leaders in whom they can believe.

How do you staff the workplace or help to improve your odds in the seniority game? This is one of many perennial problems of paint managers and one that seems continually to evade solution.

Sometimes this issue is easier to resolve in a closed, union environment than in an open employment finishing shop, although it may not seem that way to those who only view seniority as a constant frustration and reason to roll over people. But here is a way that has worked for me in such plants. It is a way to make the system really work in obtaining good people and holding them.

Start by viewing large plants as "farm" systems similar to those that feed major league baseball teams. There are all kinds of people and talent in a plant. If you were putting a ball club together, you would want to assemble a mixture of talent. Not everyone can be a spray booth specialist or paint mixer any more than everyone can be a pitcher or shortstop. You seek out specific skills by needs as you go around the plant. If you need a good equipment repairman, look for the area or crib where everything is in good condition.

When you locate the people you want, invite them to your area in finishing operations at lunch and introduce them to your needs. Two things could happen. First, they will probably be overwhelmed because someone like you recognizes their potential importance. Second, they will likely consider transferring to the finish area if you ask them to when an opening occurs. In time, they get the seniority for the job and stick there, and you strengthen the stability of your shop and improve your situation. Recognize people in this manner, seek them out, and seldom will any wage differential from job to job deter them from the change.

The value of values

I am convinced workers will be attracted to management that has sound values. They want to work for someone who believes in something of value and in whom they can believe. This is why such a philosophy is so important to attracting such people. It shows through that you care and respect them as individuals.

This means, of course, that you have to do something different from the herd. You have to be alert as you circulate about the plant. You have to be astute in your selection. It even means you may have to miss lunch once in a while to meet and talk to such people. Such recruiting has to be done with a certain style and with the recognition that the person you are approaching is also giving up his or her free time for you. It has to cut both ways.

This is a fairly simple but effective way to slowly build an organization and create *esprit de corps*. Like most great plans, it has basics that are simple to understand and execute. It is human desire and knowledge that takes bits of information and assembles them into something. The master planner can see what the assembled pieces will look like long before they are completed.

Unfortunately, in the past two decades, we in the U.S. may have defined our purposes and objectives too narrowly, too selfishly. Bread alone does not satisfy workers. We need an inspirational rebirth in leadership instead of more managing.

Never in history have people had so much and enjoyed so little. Our economy since World War II has been the most abundant in history and our standard of living the highest; yet we are more tense, frustrated, and insecure today. Industrial discontent prevalent in society today is an expression of hunger for a work life that has more meaning to it.

The trend toward bigness is resulting in the loss of individuality. The small are being swallowed by the big and the bigger, with the result that everything has been getting bigger except the individual. Those who use old management methods make the individual seem insignificant. Management feels it must perform its reorganizations and create grandiose programs to show its greatness. Yet those who know about building know the foundation must support the structure. Greatness must be supported from the very roots of the organization. New management methods do this by involving people at every level.

We have basically won the war against scarcity in our country. We have become the master of the physical world and learned how to convert our natural resources so as to conserve materials in finite supply. In so doing, we have lost the sense of being a part of that nature, and people are seeking ways to return to it.

We are learning more and more that people prefer that science be in the hands of ethical people who will use it responsibly. They don't want management to put in technical equipment to look good and cause the loss of their jobs. The people want planned integration and retraining so technology does not destroy their individual futures.

If people do not get this leadership and management from ethical persons, they begin to listen to any huckster who gets down and seeks them out. People need to have "skyhooks" from Monday through Friday. They need to feel the job is their center of life and its values are in harmony with the rest of their lives. This need for "skyhooks" applies to the boss as well if that person is to cope with the ever-increasing load of responsibility.

It is important to understand that a person's real values have subtle but inevitable ways of being communicated. These values are the intangibles, the "skyhooks," which are difficult to verbalize but easy to sense and tremendously potent in their influence.

Though seniority generally plays a larger role in the staffing of the finishing area in a large, union-influenced plant, many of the same conditions and practices prevail in smaller operations. In both cases, the key ingredient is to provide individual recognition and continue to provide it on a consistent basis.

For example: In one plant I was involved with everyone loved to play checkers at lunch. The workers would do just about anything to stop work early and eat so they could use their full lunch period for checkers. The rivalry was intense but controlled.

With everyone leaving their supporting jobs in the paint area, other production problems were created. Efforts at disciplinary action did not prove effective, and actually made many of the otherwise good employees very angry. They just couldn't understand the company's position.

Seeing this as a problem for the organization, it was my thought to approach the problem in a different manner. At my quiet suggestion, the checkers players were challenged to create a tournament. If they would do this, I told them, I would provide a trophy and some other prizes to the winners. I further offered to use the plant newsletter to highlight the event and its progress.

They all knew this would cause a lot of publicity, and people would be coming around to watch. This meant they could not leave early from their work, the area would have to be clean, and fairly soon into the tourney several would be eliminated from play. But the need to determine who was the best player was so important they were willing to abide by all these other considerations. Normally, they probably would not have done so.

It took about a month to finally play off the double elimination tourney and arrive at the eventual winner. When this was done, they had broken the early eating habit and enjoyed greater fun from the competitive effort. The same was done at other sites.

At another plant, there was a fairly strong feeling of pride around the paint area toward good housekeeping. This was fostered by having the people help in the selection of color schemes, general decor, and so forth. During the 1776- 1976 Bicentennial Celebration, they were asked to come up with some theme along the lines appropriate to the founding of our country. This effort resulted in a remarkable set of red, white, and blue drawings and artwork on the walls. Normally these people would never get to use this set of talents. The work was performed during their lunch periods and even after hours, and became a focal point of much pride. In addition, tours of the plant were always routed through the area.

On another particularly memorable occasion, I worked in a paint area with a group of men who were all members of a motorcycle gang. They were quite intimidating, and, at first, they gave me a rough time. Something had to change.

I knew such people really had one great love — their motorcycles, and they had some of the very finest. During a lunch break one day, I asked them if they would like to know more about how to paint custom jobs like candy-apple, metal flake, pearl, and flames for motorcycles. They looked at me doubtfully, but said they did. They were obviously skeptical that someone like me would know about such things.

They parked their bikes together in an area designated for motorcycles. The weather was good that time of year, so we arranged to meet at lunch each day for lessons. Together we were able to provide the basics and guide a few of them into reworking their cycles. It was a real bridge between two distinctly different types of people who had identified a common denominator.

Their barriers came down when they saw I was willing to give up my lunch period for them. Their attitudes changed, and they became more communicative. Strangely enough, this brought out some initial bitterness from their management, but this was overcome when the extra step was taken to bridge the bitterness gap between workers and their management through other projects of common interests.

Once the word spreads that management wants to understand and is willing to work with people, both within and outside the organizational structure, worker confidence rises geometrically. They want to be a part of such departments. It is not necessarily a case of money. It is simply that someone cares, is fair in his or her dealings, expects the same treatment back, stands for values, applies the rules equitably, and provides a long-term vision.

CHAPTER 7

Seed Corn, Ice Cream, and Turtles

Despite the engineering magnificence of the pyramids, they were and are in final analysis merely houses for the dead — they were designed and built to withstand and resist change. Modern paint operations — and the plants in which they reside — cannot.

The tales behind the three items in this chapter's title have been the foundation of my career and the force driving what success I've had over the years as a manager. That is why I want to share them. As mentioned earlier, one of the senses to look for in people is that of common sense. Seed corn, ice cream, and turtles illustrate lessons I learned growing up in rural Ohio that have served me well in the business world.

The seed corn philosophy

One of the fundamental lessons learned by pioneer families of old was the adage of seed corn. It was central to their survival. They knew if they were to have a crop next year, they had to hold back a part of this year's crop as seed.

The axiom has stood the test of time, and is applicable today to virtually the entire spectrum of human activity. Successful business managers do not sell their entire "crop" to enhance short-term performance. They "put something back" into the business to sustain and renew it for the future. A company's financial, technological, and human assets are all it has to function and succeed. By investing in these assets — by using its seed corn — the company stays strong and healthy to compete.

However, there still are many managers who run their business for the short term, selling or using all

55

their assets. They then move on to another area or company, claiming that they have been a high producer. But those left to salvage the operation know the real story. Down home we called that "sucking all the honey out of something."

The ice cream philosophy

It has often occurred to me that many things in life are like eating ice cream. In moderation, it is really a treat, but if we eat an entire gallon of it, we will usually get sick.

The same thing can apply to management. Managers will let things go for long periods of time, then put the press on for some reason and in a flurry of painting activity, expend great amounts of time, energy, and money to preserve a schedule. Not only does the operation suffer in terms of erratic work flow patterns, but quality is severely jeopardized by the catch-up mode of painting.

Managers who operate in such flurries are task oriented, as opposed to results oriented. They prefer to be judged by the number of things accomplished in any given period. These tasks, however, may not have been performed at the proper time or they may have been the result of an overreaction to something. Because the pretreatment line acted up, these managers may now order titration to occur every hour when it may really only require titrating four times a day.

It is important to emphasize that the right amount of time, energy, and money expended when and where it is required provides the best yields or values in your paint shop. Too many instances have been documented in which managers produced nothing more tangible than a monument to their greatness. It brings to mind the parable of the tombs in which those wielding the power prove themselves by undertaking monumental tasks. Their method is to get a block of granite large enough to meet the dimensions. They then would have to create a gigantic material handling method to transport it to the site. Next, huge numbers of artisans would have to be assembled to chisel out the rooms inside to provide the finished product. Future flexibility, of course, was nonexistent.

It doesn't matter that a lot of rubble was left from the granite. This would be removed and hauled away under the guise of good housekeeping. The accomplishment is lost in the cost of investment, expensive material handling, scrap, labor, and public relations. These monuments to success soon become granite tombs for posterity to witness, having no value-adding function.

It is far better to build an edifice with a series of little acts of leadership and good management that serve, metaphorically, like bricks. Each in itself is not very impressive, but so valuable to the end result.

The Adobe Indians, on the other hand, would have taken mud and made a quite satisfactory factory with raw materials and labor available on site. They would have made the walls thick enough for comfort inside during all seasons and not requiring much energy to operate. They wouldn't require all the material handling and other heavy capital investments. The plant would have been ready to operate much sooner, and they would have been first to the marketplace. If the business dropped off in that area, they would be able to build another plant closer to the customer, rather than be stuck with a grand, granite white elephant.

To succeed in our business, we have to avoid leaving these kinds of tangible monuments for posterity to untangle. The pyramids were great engineering feats, but they are houses for the dead. They remain unchanged over the ages. Factories and finishing operations cannot.

Once again, it is the evolutionary, flexible operation that fosters the performance that substantiates good leadership and management. Few things can remain the same if we are to be competitive. Our job is to make orderly change in the correct amount at the proper time, maintain our facilities and equipment, and provide the proper level of control. Too much ice cream only makes for a sick individual.

The turtle philosophy

We all know the story of the tortoise and the hare, so it doesn't need repeating here. But the moral is sound: slow and steady effort eventually succeeds.

If we all look about us, we can see the rabbits. They are flashy and tend to bring attention to themselves by their level of activity. But there is a hazard in these cases to confuse activity with accomplishment. Although there is a lot of hustle, bustle, and motion going on, in nearly every case, little continuity is in evidence tying these efforts together into a final product. They run out of steam and never finish the race. In the process, a lot of assets are squandered.

It is one of the characteristics of intelligent people to often not be able to remain focused on any subject long enough to complete it. They jump about, touch on a lot of people, and eventually become like a pesty insect that disturbs the flow of work, and one doesn't have to be bitten to be disturbed.

Somewhere I read that most things in life were 10% inspiration and about 90% perspiration. My experience in this business has proven this out. Perhaps this is another way of identifying leadership and management needs. The 10% inspiration would be the leadership role and the 90% perspiration would be the management effort.

This division of effort took two forms of philosophy in my career that translated into performance. First, I broke down all plans into small steady increments which people could understand and handle on a regular basis. There was a beginning and an end that would take place during some predetermined time period. I identified goals and shared my expectations. As each facet was completed, my confidence grew and success fed upon success. This builds upon the axiom mentioned earlier that management deals largely with short-term activities.

It is akin to making a rope in real life. A single strand or thread is not strong. When woven into larger strings with other strands, its strength increases. When many individual strings are woven together into a rope, exceptional strength results through synergy. The final product comes as the result of a lot of slow continuous activities taking place and gaining value each step of the way.

Secondly, I made it a point to do just one thing for sure every week while on the job, in addition to those I normally did.

For example, it might be my goal to obtain a spare paint gun or piece of equipment that would help reduce the line stoppages which would occur from time to time as we replaced parts. These spares might come through salvaging damaged equipment that would ordinarily be junked or, perhaps, if some money was saved by our efforts, a new gun might be purchased. People would see their efforts rewarded as tangible "seed corn" being put back into the business by this simple weekly effort.

The next week it might be a pledge to clean out an area of the finishing operation to improve health, safety, and housekeeping conditions. There are numerous candidates for this type of effort.

Most people come to work every day with not even a single little goal they want to accomplish. They do only what they are told. Yet, one small, but extraordinary task per week pays great dividends. One little thing a week outside the "norm," comes to about 50 goals achieved a year; in 10 years, you will have accomplished 500 things that probably would not have been done otherwise.

It doesn't take a lot of calculation to comprehend the improvement and evolutionary change just 50 accomplishments per year can make in the finishing area. My clients practice this every day. It isn't flashy, but it really does begin to pay off for those who methodically follow the parable of the turtle. Just keep moving toward your goal. You will find, as I did, that you have passed by a lot of rabbits along the way.

Such philosophy is applicable to all endeavors, work-related or not. Just like in the business world, slow and steady wins out over fast and jerky. What works in the paint shop will work at home.

Each of these philosophies assumes discipline and dedication, characteristics scarcely in evidence in an era searching for instant success and gratification. Too many people and business organizations today self-destruct from a lack of both.

To build a strong organization, what's needed is a return to the basics of success: determination, discipline, dedication, and a sound work ethic.

CHAPTER 8

Business Planning in the Paint Shop

The effective manager recognizes that efficient production hinges on careful planning and administration, and that ample time must be allocated for both.

An important ingredient in the management formula of any finishing or paint shop is what we call a business plan. In formalizing a plan for the paint shop it is important to remember that this will be part of a larger plan prepared for the company as a whole. Several things need to be taken into consideration when producing this critical blueprint of operation. It requires input from numerous people representing many disciplines and tends to go through a certain number of iterations before ultimately being accepted.

Short-term successes in business can be achieved through shrewd intuitive decisions, successful opportunism, personal ability, leadership, and drive. Increasingly, long-term success in adjusting to changing conditions is an element that must be performed if continuing growth and profitability are to be achieved. This comes about from careful strategic planning followed by diligent execution of the plan.

Repeatedly, some companies come up with the best products, provide more and better jobs, enjoy better relations with the community, and increase their profits over a long period of time. Others, which were leaders years ago, no longer hold that position or don't even exist today. What differentiates the two?

The answers lie in the fact they either had, or didn't have, leaders capable of planning their business strategy and then carrying it out.

All successful companies I have known based their business strategies on two activities: they always took a deep, searching look internally into the strengths of their business, those that could be capitalized on and the limitations that must either be reversed, overcome, or recognized in realistic planning; and they routinely investigated outside of their world to ensure that external factors affecting their business were adequately balanced with their obligations to customers, employees, owners, suppliers, and the community. Businesses who gear their policies to the dynamics of the economy, marketplace, and social trends make the best adjustments to realize opportunities for growth.

The heart of any strategic business plan is defining and setting proper objectives. As a vital part of the company, the paint operations area should integrate its plans and contributions into those of the company. Working independently toward common goals seldom generates the harmony required to perform within the organization.

In outlining the business plan, certain questions inevitably will arise. What kind of company do we want? What kind of market share do we aspire to? Do we specialize or diversify? Is growth one of the primary goals or will growth come as the result of other success on a more moderate basis?

Answers to these and other questions need to be in place and understood by the informed paint manager in order to best plan his or her contribution to a business plan.

Once a company has set its objectives regarding the type of business it wishes to be and determines the market position it wants to have, management must assess the product line's ability to meet these objectives. The size and quantity of products determine facility requirements. Facility decisions also involve considering the plans of competitors.

If you are in a new industry, it is possible for almost explosive growth in the beginning. As the market matures, there will normally be a slowing and relatively stable period, followed by declining product demand. All these impact any plan.

It is not uncommon for several companies in the same industry to look favorably at demand for their products. However, when each builds some growth potential into respective plans, overcapacity is created that may exist for years to come. When they overestimate, they end up with excess capacity and unnecessary capital investment and fixed charges. Conversely, if they underestimate, they incur the added costs of inefficiencies of overtime and other costs to meet peak demands.

Strategic planning of facilities will require a thorough understanding of the impact of automation, electronics, new technologies, competition from substitutes, and new manufacturing processes. One of the major problems besetting companies today is the mobility of junior executives who move from large companies to smaller companies. Since the only experience they have is in large volumes, they tend to prescribe a level of automation that could not really be justified with the smaller company's volume. This leads to badly needed capital for other requirements being tied up, and the creation of unnecessary higher fixed charges. With the most honorable intent, people can impose their knowledge of technology in the wrong place, in the wrong amount, and at the wrong time. In forecasting needs for equipment in the context of the paint shop's business plan, it isn't always a matter of being able to implement them, but whether they can be cost- or value-justified.

When we look at facilities, there should also be broader considerations. You need to look at the impact of decisions on employees, customers, suppliers, and the industry. If you eliminate all the employees, what will be the impact not only on them but on the community, state, and even nation?

This is why I support the evolutionary approach to finishing and paint shops. This method provides a solid and continuous improvement operation with reasonable flexibility. It permits employees to be retrained instead of eliminated, and it encourages fitting labor into a responsible attrition mode, if needed. There is no need to cut jobs in the interest of implementing new technology. Both can be accommodated by managers who are also leaders.

For example, if robots are under consideration to replace manual painting, management could select from several types of robots and various programming levels. Why not focus on a robot that could be programmed by a newly trained worker from the floor rather than one requiring an engineering background? It not only extends the knowledge of painting to a machine now applying paint, but enhances the worker as well. This is responsible improvement and management. Workers will not fear new technology when they feel it is being implemented in a responsible and caring fashion.

As mentioned earlier, one can never have too much knowledge, especially where planning is concerned. That portion of management expertise that affects future capabilities of the paint shop will determine the elements required to put together a business plan that will contribute in a significant way to the goals of the company. As the paint operation's business plan is migrated into the overall company plan, its numbers and requirements should be compatible so that little rework is needed. It should point toward the various elements required to provide a comprehensive plan of operation with a high degree of objectivity and focus.

Data-gathering and communication techniques

Certainly no plan could be created without having good data-gathering and communication techniques. These communications go in both directions, up and down, during this period of planning. Upper management is going to provide certain company direction, policies, procedures, and even forms it wishes to have you use. This is the downward flow of communication to operational people.

The upward movement will include input from both workers and operational management. These communications, whether verbal or written, should be as concise as possible, clear, and contain information to permit upper management to visualize your needs and wants quickly and comprehensively.

Experience and fine-tuning over the years will provide the skilled manager those styles and points that need special emphasis. Always give the basics of who, why, where, and how in all communications. Arrange your facts, and never write something in anger and "fire it right off" to someone. It will usually produce only negative effects, often many years later.

Always do your homework! What you might have missed in some fine point could really shortchange the total plan.

Seasonal flows and estimated annual production

You will need to project production flows and numbers for the coming year, which will require input from marketing or sales groups. This is a sensitive area and one in which I have often seen disaster strike. The danger is in being overoptimistic in expressing these numbers. Overoptimism leads to overcapitalization for market demand which doesn't materialize. This carries over to calculating potential costs or savings as well. If the sales numbers are not accurate, costs and other savings will not be realized.

It is always good to remember that manufacturing operations are normally a reaction agent to sales and the customer, not the other way around. A smart operational manager gets to know his or her counterparts in marketing and sales and nurtures a solid professional relationship with them. What this provides to the floor manager is an awareness of what is really happening in the marketplace, and that gives the edge of an earlier response period if either extreme of high or low estimates is taking place.

Someone once said that forecasting is to business what courtship is to a man or woman. While either is pursuing the person of their choice, they still keep a sharp eye for others. They may change speed and direction rather quickly or even stop running if conditions change. So it is in the business world.

In its most generic sense, forecasting is a reaction to a look at general business conditions. Specific kinds of forecasting have to do with the potential for something to occur, and involve many factors and a study of motives. Often forecasts are expressed in words and numbers on matters that are beyond the scope or control of any individual business or government.

They are really the best estimates of a situation that can be made at a particular time. The time periods may be short or very long. It is not unusual for management to talk in terms of 10-year periods or longer. One-, two-, and three-year periods are more common.

Forecasts are somewhat like a doctor's prescription. They should only be used by the person at the time for which they were made. For example, you won't want to set your next year's operation based upon the forecast for five years.

The numbers for planning production are necessary. They are part of your ability to state your needs in a common denominator, dollars. This permits both an examination or comparison to past performance and coordination with other disciplines to ascertain if it is possible to achieve company goals without undue risk.

So, we must look back to the past in order to go forward in the future. We all know that despite the wizardry of contemporary automation, there are no infallible mathematical means of seeing the future. We must couple our knowledge of what has happened before with the computerized tools at our disposal.

While all parts of a business plan are important, this one area is certainly vital to people in actual production areas. Forecasting serves as the basis of costing and budget considerations. It should be noted that forecasts and budgets are not the same thing. Forecasting is descriptive and profit planning. A budget is the plan for profitable operation of a business strategy or plan. More specifics about creation of a budget are covered in the next chapter.

Standard operating practices and audit capability

Any business planning or operation of a paint shop requires procedures or methods to handle any number of occurrences and contingencies. These are standard operating procedures and rules of doing things.

These standards become the basis of control and range from everything like passing out payroll checks to how we dispose of waste materials. It is from many of these procedures that we obtain the basis of costing and budgeting. They should be precise, specific, and unambiguous.

Anyone subject to control by others must understand clearly what is expected of them and what it means to be content in such a relationship. Without this understanding, it is impossible for the manager or operation to perform optimally.

It is possible to accomplish some task through luck or by accident, but this is not a sound basis for good management to meet company objectives. Without the use and understanding of operating procedures, people may work quite industriously while also working aimlessly and accomplishing little. Most people want to know just what they are trying to do.

Even having standard procedures in a business plan is not control. Control is the continual checking to determine if plans are being observed and progress toward objectives is being made.

There needs to be some form of auditing or checks and balances in any business planning. At least two are required in any sound plan. The first is an ability to audit the financial integrity of the operation. This is done normally through a series of budgets, cost accounting, payroll, and other tracking methods to ensure accuracy of reporting and prevention of theft. Each of us has the fiduciary responsibility to protect the assets placed under our control. The formal authority for this is usually found in the accounting area or comptroller.

The second audit should be one to ensure product integrity or quality. Quality must be manufactured into products because it cannot be inspected in. Quality is a way of life. Everything we do must be directed to obtain first-time yields. We'll discuss this at greater length in the section devoted to statistical process control (SPC) and ISO 9000.

In most paint shops, this second audit is performed by the quality organization, though of late there has been some success in getting production management and workers to take on the responsibility for quality. When you can reach this type of awareness, there is literally no need for the expense of inspectors. You can greatly help in this philosophy if you plan your facilities and flow of work in such a manner that there is a "turnaround" capability.

With turnaround, rejected paint work does not leave the originating line. The work is held and "turned around" by power and free conveyors and run back through the same process again, with the mute reminder to people on the line to do it right the first time. This is much more effective when compared to paint lines where workers are not responsible for their own rework.

It has always been perplexing to me to hear someone state, or even imply, that they know and recognize quality better than their production manager. When that occurs, there must be a great void of integrity, and there probably is cause for either training (as a minimum) or replacing the manager (as a maximum).

Up-to-date accounting systems — total costs

A solid basis for determining accurate costs of all forms must exist before any plans can be made. Probably one of the most misunderstood areas of operational management is the one of accounting. Not only do people lack understanding of accounting, often the accounting systems in place have not kept up with changing technology and changes such as waste disposal and regulatory requirements.

Terms such as return on investment, payback time, fixed costs, variable costs, direct and indirect labor, direct and indirect materials, and overhead are lost on production management.

Many of the labor and industrial accounting systems in place today are the result of the outdated method of Taylorism brought to establishing processes. From these evolved a means of tracking value-added labor and providing for money to operate support functions. As advanced technology was introduced into painting areas, these older methods of tracking and budgeting were not always in sync with more modern operations. For example, since prime piecework is not reported from an automatic operation, the accounting system cannot provide accurate reporting, unless modified.

The modern paint and finishing shop has seen a great deal of change where costs are concerned. New materials and higher transfer-efficient technology to meet environmental regulations could further impact total costs in many ways. The costs associated with all aspects of painting are seldom understood by people outside the discipline. Paint managers must make a sincere effort to educate and communicate with their fellow managers in the accounting arena.

This increased communication will result in increased levels of accuracy in estimating cost of operations and, by extension, improve profitability. We will go into greater detail of many of these elements in Chapters 11 and 13.

Engineering, maintenance, and support considerations

Personnel who operate paint and finishing shops are not normally considered to be engineers, yet, for a successful operation, many technical needs must be addressed in any business plan.

There is always a need to perform basic ongoing operational tasks for all forms of maintenance. The "pay me now or pay me later" axiom could not be more on target. If we fail to maintain equipment on a continuing basis, we end up paying more eventually in failure and replacement costs. And it seems the failure and replacement always occur when we least can afford to have a breakdown. Some shops seem to be perfect examples of Murphy's Law continuously. Others simply refuse to submit to that law and maintain and service their equipment as prescribed by the supplier.

The other engineering need in your plan involves facility modifications or creation of new facilities. Here, layout and sizing, building and utility requirements, cost quotations, and project management are the primary skills to focus on and build into a business plan.

Always remember that timing is critical in any area of the paint business. An action taken at the wrong time can be disruptive and costly in terms of penalties to a business. Having the plan and the money to carry it through does not necessarily guarantee success. As important as money is to a business plan, it can be money poorly spent if schedules are overlooked.

Use of time

Time is an asset that must be understood and included in planning. We use it to measure how much can be performed. We pay our workers based on it in most cases. Once wasted, we can never get it back. It can be wasted by doing nothing or by doing something wrong. Sometimes it flies by, and sometimes it is as though the hands are nailed down on the clock. We do not know how much of it is allotted to us in life, and there is no guarantee the elements in nature will cooperate when we want to do something in particular. Why does it always seem to rain when we want to pour concrete or go on a picnic?

Good managers recognize time as a valuable planning tool and asset. They create goals, timetables, and set priorities so as to manage time effectively.

They also create alternatives so that if one activity is blocked for reasons beyond their immediate control they can substitute another part of their total plan. Just as the little bricks we mentioned earlier lead to the building of a functional edifice, little actions end up being a part of the finished product, on time, and within budget.

The effective manager realizes time will be needed not only for production, but for the planning and administrative demands as well. Reports must be routinely prepared in any operation. These must be completed on time and in proper form so that other disciplines in the organization can accomplish their work. It is essential. Failure to do these things and allow for that function in your business plan is a strategic error.

Policies and procedures

If the proper forms are created and made as user friendly as possible, less time will be needed to report on activities. The use of computers and other electronic means of communications may be the solution to this need. Whatever the way, put it into your business plan.

Information systems should relieve management of routine tasks, open up time for analyzing problems, facilitate planning, and increase performance. This technology can take input from disparate groups well outside the paint shop. While working on problems independently, you may often find these efforts turn out to be closely related and even cross-fertilizing. I have found it most helpful in the development of techniques for conceptualizing and measuring information. The use of simulation provides the capability to actually see if bottlenecks or other problems exist when planning any major changes in either facilities or production scheduling.

Strategies can be analyzed and tactics honed and proven in various "think" games. One of the potential problems with the use of such electronic information systems is one relating to social and psychological impacts on people. There is often a fear produced in people concerning how vast amounts of information are being used.

People are concerned about the systematic manipulation of information today on individuals, groups, or even machines. Questions like Why do you need it and what are you going to do with it? are quite common. Many feel this information will somehow be used against them. This concern must be erased, or it can lead to poor morale, reduced performance, and the loss of an important tool.

Manpower/labor

The use of human resources will be determined usually in two manners. The first will come as a result of specific numbers generated from the forecasting conducted on schedules required for production and supporting services such as maintenance and technical/engineering services.

The other usually comes from such areas as replacements for sickness, vacation, death, pregnancy, and training. Most firms develop statistics for what is normal and apply that percentage within business planning and budgetary figures.

These numbers can often tell us a great deal about one operation *vis a vis* others in direct competition or from industry to industry. They serve as a valuable tool to determine how you stack up. Are yours higher or lower than normal? Do your numbers reflect some real problem of morale, safety, working conditions, or other factors that influence why employees miss work? Are your turnover rates for employees higher than others? (This can have a dramatic impact on costs to retrain and orient people, as well as rework costs due to lower quality attendance to new worker ramp-up.)

So we have both direct and indirect type labor to deal with. Over the years, the ratio of direct to indirect labor has been changing because of new technology emerging in the paint shop. The increased use of automatic equipment and other process improvements have reduced the need for direct labor and put more emphasis on support personnel.

As mentioned, accounting systems must keep pace with present-day needs. If your accounting system is based on direct labor or incentive piecework labor in order to generate a certain amount of support labor, then you may find yourself being pressured to reduce the maintenance or other activities as technology replaces direct labor. This is a sure sign something is out of line. There must be new ways of providing for support labor in highly automated paint shops and reasonable allowances where automation is in the implementation stage.

This requirement was borne out in a case in Europe several years ago. A new, sophisticated finishing line was installed that actually took no direct labor to operate. The only personnel required were support people. On paper in the accounting system, this paint shop did not exist or provide any value improvement because no prime labor was produced by people.

The system of accounting was not changed to permit the use of such technology. The finishing system was altered to use less technology and include enough labor in the loading and unloading areas so that production could be tracked. The ratio of direct to indirect was revised. Every day the line operated, it was an improvement to what had been performed previously, but it did not provide all the capability it could have (which may have been a contributing factor in this business being sold).

Training/educational requirements

No successful business can exist in a competitive environment today without consideration for increasing its human resource's level of knowledge. People today must know more about all facets of the operation and be adaptable to new methods and technology. We do this through effective communication and training.

This need can be fulfilled in several ways, both internally and externally. Whether your business plan provides for a huge in-house effort or some combination, it should be an integral part of your plan and funded accordingly.

Training can further be divided into self-training and structured training. There will be a number of positions in the organization that require a formal education and degree in a respective field. It has only been in the last decade or so that it is possible to study for a formal degree in paint technology. Most people involved in the management of a paint shop today are not degreed in paint technology. Their's is an on-the-job education, which optimally includes some basics of chemistry or manufacturing engineering.

Self-training has played a vital role in my career. I was encouraged to read, build models, tear down and rebuild junk things, and become innovative to pass the time. I learned how to make crystal radios and tear down and assemble small gasoline engines. We all knew how to repair our own bicycles if we had them. All this formed the basis for being inquisitive and served as the earliest form of my education.

Later in life it was customary to get a job and try to get a more formal education while earning a living. You were lucky if you could find a company that might help pay your tuition. I was fortunate in that respect, and used the system to work and earn a degree over a 10-year period.

Before and since, I have used every opportunity available to obtain knowledge where and when it was possible. Such efforts train and hone the discipline necessary to plan and carry out goals. The pride of performance will overcome stress and many other physical problems that can plague people when they are not satisfied with their careers and lives. All these things are also part of the little bricks they build their professional and personal houses with. I rigorously support self-improvement efforts.

The Society of Manufacturing Engineers offers opportunities for obtaining both a technologist level and full certified manufacturing engineer level certification in the area of finishing and coatings. It is recommended that paint and finishing management take advantage of either of these programs as a means of increasing their knowledge and that of their staff.

As I alluded to earlier, paint people 40 years ago had a terrible image. They were not considered professional in any way, and they earned a reputation for hard living and being generally profane. Further, various transactions between suppliers and management were replete with behavior now deemed improper if not illegal. All the wrong tactics were often employed to obtain business.

Moreover, since it was against the rules to smoke in paint areas, many people chewed tobacco or used snuff. The residue of such habits provided upper management with a perception that was quite negative. Professionalism, or even the perception of professionalism, never graced the paint shop floor.

As a result, upper management formed a self-fulfilling prophesy in many cases. They viewed paint people with suspicion and withheld funds and talent from the paint areas. They also failed to train and educate the people. With this lack of support, the paint shops were doomed to playing catch-up with the other divisions.

Investing in people is more important from a productivity standpoint than acquiring new facilities and other assets. It is through training and work force development in our business planning that we gain continuous improvement and profitability. This is an investment in better working conditions, a means to increase wages, and improve benefits. This is not entitlement. This is getting results the old-fashioned way: by earning them.

If you are failing to obtain the expected results in your paint operations, it is the result of not planning and committing the necessary effort and resources.

Direct/indirect materials

No business plan can be complete without the inclusion of sound cost data or estimates of materials required. These are normally carried in two categories. The first are the direct materials used in producing something. In the paint shop these are, of course, the paints, solvents, sealers, and products that actually leave the plant on the finished product and shipped to a customer.

The second, or indirect materials, are those things used somewhere in the processes, but do not leave with the product. Examples of these would be masking tape, tack rags, repair parts, uniforms, gloves, and maintenance items. In some accounting systems, you may also find the cleaning and phosphating chemicals listed as indirect materials. The same applies to chemicals used to detackify paint overspray and other water-treatment chemicals.

In addition, some systems might include such items as energy costs under the general category of materials. Others may list them as a separate distinct line item for better visibility and tracking. Once again, form doesn't matter as much as being sure it is included. The important thing is that the more line items you can identify and place in the accountability of management in a paint shop, the more visibility and accountability you will have in trying to control and improve them.

When people can see the results of their efforts to save materials, energy, or any part of costs directly showing up in their accountabilities, it becomes far more personal and motivates them to perform.

Surprisingly, some major companies' accounting systems do not charge paint and other direct materials to the budgets of paint shops. They place them into a group considered "bulk materials." When this occurs, you will generally find low attention levels paid to reducing the paint bill, since it is not one of the paint shop's accountabilities. When given a choice to add labor to reduce materials, you will almost always find management opting for wasting materials to save labor.

Today that practice is deadly, for a variety of reasons. First, the cost of materials has increased faster than the cost of labor. This is due to many of them being tied in some ways to oil or energy used as their base stock. Next, increased uses of materials increases VOCs and potential waste disposal charges and problems. It can impact health, safety, and other workplace concerns.

Materials have become the major portion of expense today in most paint shops. Taken in the context of the number of changes these materials will undergo to satisfy environmental requirements alone, there are sure to be increased costs ahead for the paint line. These have to be overcome through better utilization and elimination of rework and waste in the future.

This now has a direct bearing on facility modifications and different methods of operation. You cannot go through a business plan without observing how the different parts play into one another. Certainly materials, both direct and indirect, are major players today in a healthy business. Economically, they impact the paint shop's efficiency; socially, they influence the way managers interact with their staff and their peers.

Overhead

Finally, there will be an item attached to any plan broadly categorized as overhead, sometimes called "burden." A simplistic way to describe overhead is that it includes all costs other than raw materials and labor associated with the manufacture of the products. Based on the accounting system employed, indirect labor mentioned earlier could be included as overhead. Other labor such as supervisors, transportation, janitors, and office personnel also could be included. General and administrative costs may or may not be identified separately and included.

Manufacturing overhead also includes such costs as heat, light, power, maintenance, and supplies; and depreciation, taxes, and insurance on assets. Overhead may be calculated and charged uniformly as a total plant or broken down into departmental operations.

There can be fixed overhead that changes little regardless of what is being produced. Insurance on a building is one example.

Other overhead items can be variable, changing depending on level of production. If you produced 200,000 of something instead of 100,000, the overall cost per unit would go down.

Still other overhead is directed at a given product and can be variable. If the product line were removed from production, those costs directly associated with tooling, and so forth. would be calculated differently than those of another product. The point is, overhead allocation can vary from business to business, but it must always be a part of any business plan. In most cases, this allocation is calculated on an annual basis. The rate is changed at times if some volatile situation occurs. When that happens, the figures used are estimates of costs rather than historical records of what they actually are.

Some costs are classified as overhead because it is impossible to associate them directly with products. Other costs are classified as overhead because it is not convenient to trace them directly to products, even though it would be possible to do so. Nevertheless, total overhead is properly part of the cost of total products worked on, and some reasonable part of total overhead must therefore be charged to each unit of product.

In summary

It is important that each year some form of accurate business plan be prepared in any organization. It must be a total plan for all disciplines, and the paint shop like all other departments must contribute to the plan.

This will make it necessary to obtain data from a wide variety of other disciplines that affect the paint shop in order to present the most accurate and timely plan. Business plans require good communication. Production flows and total numbers, user friendly procedures, forms, and methods to list and report data must be carefully thought out and included. An understanding of the accounting system is a must, as are engineering, maintenance, and support considerations. A workable plan assumes an ability to manage time and training, and sets out provisions for education, labor, materials, and overhead considerations. To this must be provided the factors of timely implementation and integrity of accountability for performance.

Business plans can take various formats based upon individual companies' management styles and accounting systems. Experience has shown that paint management must be at the forefront with its input into the overall plan in order to obtain the budget and function at the best levels of performance. The *total* manager understands, speaks, and performs using the business tenets and language of the paint shop.

CHAPTER 9

Budget Planning

An in-depth understanding of costs and the capacity to respond through a system of continuous improvement set the effective manager ahead of his or her mediocre counterpart.

All businesses make plans. They have to. These plans tell us what the leaders are thinking, what the objectives are, and hopefully, the best way to reach those objectives.

A variety of approaches

People approach the planning task from a variety of directions. Some plan entirely in their heads. Others make notes on anything available to write on. Still others make highly detailed plans in an orderly, systematic manner.

It is the latter group we will examine because it best fits what occurs in most businesses. We will call this plan a budget, as most items are expressed in a quantitative form. A budget is merely a plan usually expressed in monetary form. Budgets also can be expressed in forms such as unit of product, employees, or units of time. But whatever expression, budgets are a tool in business for both control and coordination. Because the paint shop is a business almost within a business, it falls under the purview of budget constraints as well.

All businesses have some form of budget, and several types abound. Operating budgets show planned operations for a forthcoming period. Cash budgets show the anticipated sources and uses of cash. Capital budgets show planned changes in fixed assets. Paint managers should have at least an

77

understanding of these various forms of budgets. Because many smaller businesses do not always have all of these budget forms, our focus will be on the operating budget, as most departments and businesses do have these.

Operating budgets can consist of two parts. One, called the "program" budget, describes the major programs the company plans to undertake. It may be aligned by product lines and show the anticipated revenue and costs associated with each product. Executives tend to use such budget forms more than operating personnel and middle management. They address such issues as whether profit margins per product are satisfactory or balanced with others and if production capacity is in sync with the capability of the sales organization. Program budget planning also ensures that cash funds are available for the organization's needs, and continuous review of the plan ensures that questions arising about any of these issues get the attention needed.

Most of management are exposed to the "responsibility" budget, which sets forth plans in terms of the person responsible for carrying them out. It is a control device in that it is a statement of expected or standard performance against which actual performance can be compared.

The principles of budgeting apply in that budgets should be sponsored by management and must be regarded primarily as tools of management rather than accounting devices. The costs shown should reflect separately the controllable costs of each responsibility area.

The architecture of budgets

Responsible managers are an integral part of the budget process because their input is essential. Only they can supply the real shop-floor numbers on which to form a budget. And since they are charged with meeting the subsequent budget, their participation in the process becomes central to assessing alternatives and setting budget goals. That is why it is important that managers have a thorough understanding of what the budget process entails. Staying abreast requires an ongoing effort of reading manuals and memos, and participating in brainstorming sessions to discuss new budgets in terms of actual results achieved with current ones.

A time period for the budget is a must. Among other things, it enables management to assess realistically how effective the company is in reaching its goals. Usually the overall time period for a budget is one year, subdivided into weekly or monthly periods. A weekly budget is perhaps preferable because it permits closer scrutiny and faster reaction to any area beginning to drift.

Budgets are now being made paperless in many plants as the use of electronic communication is instituted. To some extent this can lessen the visibility to the overall organization depending on the level of computer use throughout the plant. In the case of low use, budget data becomes hidden in the computer.

One element in budget recording that seems perpetually to be surrounded by confusion is that of actual costs versus recorded costs. Most budgets have an area for reporting "direct labor," but these only reflect actual wages paid. They seldom list the other costs such as vacation pay, social security, retirement, medical, and other fringe benefits. Although these costs are accounted for, they don't appear as part of total costs of labor. If they did workers would have a better understanding of just what they were really earning.

A budget should have or represent goals that are reasonably attainable. They should not be too tight and cause frustration nor too loose to encourage complacency.

Line organization people should make the plans and prepare the budget with staff people assisting. Thorough reviews on a regular basis should be made by ever-higher levels of management. This not only exposes the plan to greater scrutiny, but sends the message through all ranks that executive-level management is interested in all activities of the company.

Approval or disapproval of both budget and performance should be specific and communicated to the organization. A policy of "silence gives consent" will inevitably lead to poor communication, mistaken understandings, and poor performance and morale.

Integrity is critical

No budget is any better than the quality of information provided and how the information is used. The personal motives of some managers can greatly influence the manner in which a budget is planned and how their performance will be evaluated. Unfortunately, this leads to some becoming masters at using the way a company handles distribution of overhead and other costs to their personal advantage. Responsible managers should be alert to them.

To illustrate, consider this example. Joe is a "lean and mean" type of manager, very careful about not increasing costs in his department. Harry, on the other hand, could be called a "fat cat," and is just the opposite. Because fixed costs in their company are not allocated to the division, Harry encouraged their expansion. When headquarters finally noticed a drastic increase in the fixed costs of the entire organization, it ordered a 15% reduction in these costs for every department.

Harry easily conformed (even beat the target) while the honest "lean and mean" manager failed. The report went to corporate on the performance of both. Harry was promoted and Joe was fired for lack of performance.

This scenario has been repeated with different twists many times. Often promotion or discharge does not occur, but raises have been given and raises withheld. This is indeed sad because it maligns the budgetary process. Often it leads to such practices as purposely overstating initial budget requests with the knowledge that past practice has been to reduce those estimates by a certain percentage. These practices lead to companies not knowing or being able to effectively control costs and performance.

Thus, a direct costing system that eliminates the allocation of costs to individual departments creates a dual problem. First, individuals operating in their own self-interest perceive no incentives to reduce fixed costs but, instead, see incentives to increase them. Second, information is distorted because managers will inefficiently substitute fixed costs for variable costs while appearing to be more efficient.

Another area vulnerable to budget manipulation is depreciation. Depreciation is normally treated as a fixed cost based on the cost of equipment. Managers wanting to skirt the process continually press for more and larger machines than they really need. The more the equipment costs, the more is budgeted for depreciation.

The important thing is that the responsible manager be ever vigilant in the budget process for these types of abuses. No cost accounting system can measure adequately which individual units of an organization have their costs entirely under control.

A carefully prepared budget will serve as an excellent communication device. It will become a two-way commitment between the upper management and the operating supervisor. It serves as well as a standard of measurement for actual performance.

The element of flexibility

The final element that must be part of any good budgeting process is one that recognizes "exceptional situations" when they occur after a plan has been placed in operation. Implicit in the plan should be the recognition that events can happen that are outside the bounds of the original document, and adjustments will be made when evaluating performance. This level of flexibility is imperative in dynamic organizations and underscores the need for strong manager participation.

Good management will have a thorough understanding of costs and the ability to react through a system of continuous improvement. At the beginning of this book, it was stated that it is difficult to know where one subject breaks off and spreads into another area of the management activity. Arguments are made that costs should be the sole domain of the accounting department or part of the quality and continuous improvement effort. Costs are a budget item and should be included in the budgeting area. Accounting does not create or reduce costs. It tracks them. It is my feeling that monitoring our systems for cost control comes after identifying and establishing performance standards. Calculating realistic estimates of costs is vital to an accurate budget and reinforces the other activities in a well run paint shop.

Budget "adjustment" is another area that commands management attention. It is quite possible in a business enterprise for a manager to be granted increases in allowances for all the wrong reasons. A set of unusual events may have characterized the paint shop during the preceding year, requiring that adjustments be made. If the budget process does not recognize these, it is possible to use those exaggerated figures as the basis for this year's budget. Over the years, as much as 30% to 50% waste and other inefficiencies may become built into standard costs. Workers and managers alike, in meeting their budgets every day, really believe they are operating an efficient shop. If the standard is wrong, then the performance is wrong.

Activity-based costing

A new concept finding increasing acceptance in the management area is that of Activity-based Costing (ABC), a part of Activity-based Management. Rising out of the need for more accurate ways of identifying and assigning costs, ABC is seen by many budgetary experts as the cure for outmoded and failing systems of cost accounting.

Labor costs today are shrinking in many paint shops as compared with post-World War II periods, necessitating more accurate ways of assigning overhead costs to products. In traditional methods, any cost objective shares in all the costs necessary for its existence. This is a reasonably fair way if the base used is correlated to overhead. But when volume-driven burden rates are used, low-volume products tend to be undercosted, while high-volume products tend to be overcosted.

Activity-based costing attempts to identify the reason for the existence of an overhead item. For example, under a traditional purchasing system for paint materials, there may be $100,000 in department costs to purchase $1,000,000 of paint. Typically, an order for $5,000 worth of paint would require the same amount of time and effort to purchase as an order for $1,000.

In most current systems of costing, the $5,000 order would be charged with five times as much purchasing overhead as the product costing $1,000. It is the action of issuing a purchase order that drives the cost and should be the base for allocation rather than the dollar value of the material.

Take, for example, special-order paints. They get the advantage of shared costs with the high-volume standard paints, which tends to make the special business look inviting and profitable because it is being subsidized by the standard business. On the other hand, the bulk of the business is being penalized due to the improper allocation. This often leads to poor decisions both in marketing and manufacturing.

Activity-based costing methods allocate costs more equitably, but adoption of such a concept should be done with care. The implementation of any new approach can be extremely time-consuming and expensive. Transition involves extensive work with numerous records and personnel. A company's very culture must change to view costs from a cost driver point of view rather than from only a labor point of view. In addition, it may require a host of computer programs to track information and costs differently.

A pilot study is recommended to test the method against any present system. If the results are not significantly different than the current system, then ABC is probably not worth the cost of implementation and maintenance. What works for one company may not for another; however, there is a need to review most accounting systems for relevancy to create a sound paint shop budget around them.

Activity-based costing would appear to be a tool that modern management may want to look into as it pursues efforts to improve performance on a continuing basis. Figure 9-1 indicates how ABC supports a wide range of decisions within the total business organization.

Departmental budgeting

For those who may have never seen or worked with a departmental budget, refer to Figure 9-2, an outline of a fairly inclusive budget. It covers the primary areas of direct and indirect labor, and direct and indirect materials, plus a number of other cost items. In this example, items such as recognizing inexperienced personnel, unique operating conditions, retroactive pay from a unique negotiation settlement, time off for physical examinations and

Budget Planning

Activity-Based Costing Supports a Wide Range of Decisions

Periodic (Strategic)	On-going (Operational)	Day-to-day (Tactical)
Competitive Evaluation • Benchmarking • Strategic Make/Buy Market Strategy • Joint Venture/Teaming • Product Line Rationalization Facility Planing • Facility Modernization • Facility Rationalization Organizational Restructuring	Business Planning Product Development • Life Cycle Costing • Design to Cost • Concurrent Engineering Capital Investment Capacity Management Pricing	Cost Tracking and Monitoring • Forecasting • Budgeting • Performance Measurement • Investment Management Continuous Improvement • Cost • Customer Service • Lead Time • Quality

Figure 9-1. Though more equitable in assigning costs, activity-based costing methods should be considered only after benefits over current systems have been proven.

Managing a Paint Shop

XYZ Company   **Weekly Operating Statement**   Week No. 49   Week Ending 9-11-93

Department: Body Paint   Gen. Foreman: L. Smith   Supt: A. Jones

Current Week Columns

Fixed Amount	Var. %	Planned Variance	Total Allowed	Actual Amount	Variance Amount	Account Number	Account Description
				3,429		00	Base prime labor
3,429	776.997		3,429	3,429	2,471		Prime lab developed
	776.997		26,643	29,114	2,471	20	Hourly increase adj for 01
3,429			30,072	32,543			Total conversion labor
378	44.048	145-	1,365	908-	2,273	03	Excess cost of untimed mfg
	7.505		635	916	281-	08	Allow for inexperience
422	141.896	422-	4,866	1,834	3,032	09	Allow for operating cond
1,836	39,212	1,279-	1,902	2,481	579-	11	Rework and salvage
2,636	232.661	1,846-	8,768	4,323	4,445		Total excess conversion labor
938			938	840	98	06	Material movement
				83-	83	07	Quality control/inspection
						14-02	Retro pay for current year
938			938	757	181		Total sup serv - other labor
350			350		350	31	Group leaders
1,024	13.782		1,497	1,745	248-	34	Janitors
1,374	13.782		1,847	1,745	102		Total sup ser - operating labor
86	4.897		254		254	42	R & M bldg and bldg equipment
86	4.897		254		254		Total sup service - indirect labor
			160	160		97	Labor union expense
	4.658			280	280-	97-10	Invest and disc griev w/mgmt
	4.658		160	280	120-		Total sup ser - 1 union - labor
2,398	23.337		3,199	2,782	417		Grand total supporting services
454			454	11	443	10	O/time prem - wage earners
40	24.120		867		867	12	Shift premium - wage earners
494	24.120		1,321	11	1,310		Total prem - wage earners
16	.197		23		23	43	R & M mach and aux equip
286	3.596		409	331	78	47	Cutting tools
17	.219		25		25	48	Abrasives
1,157	14.533		1,655	2,612	957-	49	Supplies
1,476	18.545		2,112	2,943	831-		Total indirect materials

84

3,429	1.569		54	25			Material scrapped - purch
7,004	1.569		54	25		161-23	Total scrapped materials
	2.923		100		29		Bereave hguard jury mil enc
	2.923		100		29	72	Total employee insurance, tax, etc.
	301.586	1,846-	3,429		100	212-01	
	776.997		15,500	10,059	100	213	
	1.569		26,643	29,114	5,441	213-20	
			54	25	2,471-	7563-01	
					29		
				93%			Budget efficiency percent
100			100	100		191-06	Serv received from tool room
301			301	301		191-76	Serv received from tool room
38			38	38		191-77	Serv received from machine repair
123			123	123		191-80	Serv received from electrical
121			121	121		191-81	Serv received from steamfitter
997			997	997		191-82	Serv received from millwrights
49			49	49		191-83	Serv received from garage
1,729			1,729	1,729			Total services received
12,162	1080.15	1,846-	47,355	44,356	2,999		Grand total of budgeted accounts
				94%			Budget efficiency percent
				5,143		243-A	Vacation pay liability
				813		243-A1	Vacation pay liability
				150		290	Vacation and Christmas B Hourly lia
				6,106			Total non-budget accounts
				1,087		195-01	Units produced
				39.22			Cost-P-Chassis 211-212-213

Figure 9-2. A comprehensive departmental budget recognizes all cost items and is sufficiently flexible to accommodate extraordinary items.

Managing a Paint Shop

Account Number	Account Description	Fixed Amount	Cumulative Columns Planned Variance	Total Allowed	Actual Amount	Variance Amount
00	Base prime labor				89,400	
01	Prime lab developed	89,398		89,398	89,400	2-
20	Hourly increase adj for 01		1,959-	692,658	717,956	25,298-
	Total conversion labor	89,398	1,959-	782,056	807,356	25,300-
03	Excess cost of untimed mfg		5,364-	32,603	47,988-	80,591
08	Allow for inexperience	8,724	135-	14,963	14,342	621
09	Allow for operating cond	9,898	8,855	123,392	30,188	93,204
11	Rework and salvage	43,104	31,127-	45,798	39,326	6,472
11-01	Rework - obsolete and surplus			35		35
11-02	Rework - engineering	35		952		952
18	Scrap loss - labor	952			63	63-
	Total excess conversion labor	62,713	45,481-	217,743	35,931	181,812
06	Material movement	21,673	985	22,658	23,480	822-
14-02	Retro pay for current year				556	556-
	Total sup serv - other labor	21,673	985	22,658	24,036	1,378-
31	Group leaders	8,092	229	8,321	9,137	816-
34	Janitors	23,637	319	35,652	29,153	6,499
36	Other operating labor		177	177		177
36-02	Fire call and fire drill				18	18-
	Total sup ser - operating labor	31,729	725	44,150	38,308	5,842
42	R & M bldg and bldg equip	1,996	212	6,372	2,210	4,162
	Total sup service - indirect labor	1,996	212	6,372	2,210	4,162
97	Labor union expense		84-	3,871	7,122	3,871
97-10	Invest and disc griev w/mgmt		191	191	7,122	6,931-
	Total sup ser - 1 union - labor		107	4,062		3,060-
16	Job order labor - non cap	763		763	762	1
	Total sup ser - J/O - N/C labor	763		763	762	1
	Grand total supporting services	56,161	2,029	78,005	72,438	5,567
10	O/time prem - wage earners	10,519	18	10,537	1,013	9,524
12	Shift prem - wage earners	976	473-	22,065	3	22,062
	Total prem - wage earners	11,495	455-	32,602	1,016	31,586
7548	Materials used in tests				7-	
	****Descrip not in prog***				7-	
43	R & M mach and aux equip	391	5	573	1,032	459-
47	Cutting tools	6,979	71-	10,123	5,652	4,471
48	Abrasives	416	4-	608		608
49	Supplies	28,232	173-	41,050	32,991	8,059

Budget Planning

Code	Description	Col1	Col2	Col3	Col4
7854	Total scrapped materials		1,401	2,264	863-
	Retro-wage and salaries paid			44-	
	****Descrip not in prog**			44-	
72	Bereav hguard jury mil enc	53-	2,434	2,462	2,434
72-01	Bereavement	427	427	861	2,035-
72-04	Military				861-
	Total employee insurance, tax, etc.	374	2,861	3,323	462-
7857	Salaries abs-top management			27-	
	****Descrip not in prog***			27-	
212-01			89,398	89,400	
213		43,776-	89,398	152,383	231,182
213-20		1,959-	383,565	717,956	25,298-
7548			692,658	7-	
7563-01			1,401	1,769	368-
	Budget efficiency percent			82%	
191-06	Service received from tool room		164	164	
191-07	Service received from machine repair		24	24	
191-52	Service received from product insp		163	163	
191-40	Service received from electrical		23	23	
191-76	Service received from tool room		8,755	8,755	
191-77	Service received from machine repair		798	798	
191-80	Service received from electrical		3,392	3,392	
191-81	Service received from steamfitter		4,359	4,359	
191-82	Service received from millwrights		21,561	21,561	
191-83	Service received from garage		1,287	1,287	
	Total services received		40,526	40,526	
	Grand total of budgeted accounts	296,311	45,735-	1,207,548	1,002,451	205,019
	Budget efficiency percent			83%	
193	Job order charge outs			762-	
243-A	Vacation pay liability		151,301		
243-A1	Vacation pay liability		36,740		
251	Holiday and Christmas shut pay lia		44,148		
282	Illness or personal absence lia		1,343		
290	Vacation and Christmas B hourly lia		10,325		
	Total non-budget accounts		243,857		
195-01	Units produced		29,193		
	Cost-P-Chassis 211, 212, 213		32.81		

Figure 9-2. (Continued.)

medical treatment, and bereavement are included to give a more precise cost reading to the manager.

The chart gives a fair accounting of what it costs to carry someone on the payroll. In addition to actual wages, fringe benefits of insurance, vacations, workmen's compensation, and pension costs drive up actual employee compensation costs by nearly one-third, and these costs continue whether the employee is at work or not. While this budget has more line items than some, it still does not include all types of costs.

The terminology used in budgets should also be explained for those who are unfamiliar with such activities. Direct labor is almost universally accepted to mean the specific labor involved in producing something. In the paint shop, it would be the sanders and painters, for example. It is always the labor required for a piecework operation. Indirect labor is the labor supporting these operations, such as the material movement or paint mixing personnel. In most accounting systems, a predetermined ratio is applied to the number of prime or direct labor to establish the number of indirect support labor.

Direct materials normally refer to those materials used to produce what the customer ultimately receives. Indirect materials are those products used to assist production. Items such as sandpaper and tape would be used in the paint shop, but they would not be shipped as part of the finished job to the customer; hence, they are classified indirect.

Items such as spare parts to repair equipment are usually listed in another category. They are considered to be *burden* parts or *P & M* for production and maintenance needs. A plethora of accounting rules and practices resides in plants throughout industry for categorizing the same items in different manners.

For example, in some operations, such as pretreatment, chemicals fall under burden costs. In other places, they are considered to be indirect materials. Some will argue that the customer does not receive these materials since they are used to clean and prepare the product for primer or paint. This is true for most of the chemicals involved; they are used and then rinsed off. But what of the phosphate or the final sealers used? They stay on the metal.

In some industries, it is common to "blend" short-order special-color needs in-house from basic materials rather than order them as finished products from the supplier. A bucket is involved in either case to transport the paint. In most accounting systems and budgets, the bucket is included in the price of the paint, and the cost of the bucket becomes a part of the direct material. If the plant were to blend the paint, then that plant must furnish the bucket as part of the operation. This same bucket, serving the same purpose, is now considered a burden expense item just as rags, gloves, and toilet paper are.

This begs the question, "So what? Cost is cost in the final analogy." Quite true, but the needs of cost accountability are not entirely met, and budget manipulation often results. It comes down to who is being charged the expense. If a plant has a lab person controlling the washer/pretreatment system, that department will have the funds in its budget and not the paint shop. This person often works in an area removed from the production department, and there could be conflict because the production people have no direct accountability to reduce soils or even schedule product in the best interest of chemical use. Maintenance of those systems is in many plants the responsibility of still another department that answers to neither the technical labor area nor the production department. So, often the timely maintenance and dropping of tanks in various stages is geared to another budget. Three budgets could be involved with three different managers who each may march to a different beat.

It's not uncommon in some plants for the maintenance people to raid areas and obtain supplies in order to look good in certain accounts. The same is true for equipment. Many a production manager has had delays in getting started due to "midnight requisitions" having depleted the supply of materials he or she needs.

It is for these and other instances of chicanery prevalent in paint shops that maintaining a well managed shop is essential. Production, maintenance, and other supporting activities affecting the performance should be placed under the manager of paint, along with the budgets.

Preparing a budget, then, goes beyond just accepting the status quo of many organization charts. Good budgeting begins with an organizational *philosophy* that will permit ready identification of activities and needs in a comprehensive manner. In this way, a good manager will be able to acquire the appropriate amount of funds and personnel, without overlap or need to fall hostage to other supporting disciplines.

In a well run shop, managers are not forced to play the gamesmanship of budget monopoly. Effective paint managers display the willingness and capability to accept responsibility for all the operational and budgetary functions required to run an efficient paint operation. Moreover, supportive upper management will recognize the wisdom of this approach and take note of the better performance and accountability.

The budget as bellwether

The importance of careful budgetary planning cannot be overstated. Budgets are used for planning and control, and their underlying principles are as cerebral as they are tangible. Budgets should be established in an order of setting guidelines, preparation, coordination with others, review, and, finally, approval. When in place, the budget becomes the driver of performance.

The budget is an agreement or contract between the paint manager and the company. It should command the attention and subsequent efforts on the part of the entire paint operation staff to meet it or come in under it. Positive budgeting performance is the manifestation of personal and professional integrity that results in recognition of a job well done.

The budget is also a powerful indicator of what your company is looking to accomplish and the direction and content of its goals. Understanding and learning how to use and interpret it on a continuing basis is a must for good paint management. It serves as the foundation of continuing improvement in everything surrounding a paint operation. It is incumbent upon those who do not know the business side of paint to become familiar with how budgets impact day-to-day operations. Understanding both the operational side and business side will elevate performance in a dramatic way, and your staff will view the entire operation from a new perspective.

CHAPTER 10

Planning Facilities and Painting Strategies

The single reason for being of the paint shop is to help the company as a whole reach its strategic goals in a safe and efficient manner.

Critical to any evolutionary management approach to painting are the planning, design, and posture of the operation from a facility point of view. With such an approach, flexibility is key. Capital investment should be carefully planned for flexibility so it will have as extended a life as possible. This is normally made possible by (1) good maintenance programs and (2) designed-in flexibility to make minor modifications from time to time without interfering with normal production.

Basic design considerations for a finishing line appear in Figure 2-1 of Chapter 2. We will discuss this chart area by area to broaden understanding of the options available to provide the flexibility needed for future strategies through facility planning.

Basic design considerations for a finishing line

Reason for coating
• Appearance
• Corrosion protection
• Handling/surface protection
• Insulation/conductivity
• Machining

When a company decides to make a product, decisions will be made about whether it will require paint. These decisions will include a variety of considerations ranging from cost to durability and how the product is to be manufactured. What kind of appearance is expected? Does the product need a smooth, hard, glossy finish? Will it require a mono-layered finish or one that is base/clear? Will it have a textured finish? Will it be textured from the paint application method or will the substrate contain the texturing? Why does it need any finishing at all? These are a few of the questions that begin to define appearance.

In addition, some consideration will have to be given to corrosion protection, because such treatment will impact selection of substrate, potential tooling lubricants, cleaning and pretreatment, primers, racking, assembly procedure, application considerations, and curing needs. Readily apparent is that a single decision to paint or not paint begins to explode into a broad range of other categories of consideration.

Does the finish require any special kind of handling and surface protection — either during manufacture or while painting — such as special grounding for a nonconductive substrate? Will the parts be painted separately, or will it be possible to do several at a time in a custom racking device? (It may be possible to pretreat them that way, but not to paint them.) Must the product be dried or can it be processed wet or damp? Does the product require priming as a short-term protection and handling aid while being shipped elsewhere for painting? These, too, are issues of major concern to the paint facility planner.

Some products, such as those made of plastic or fiberglass, are coated to provide conductivity. Others, principally in the electronics industry, are insulation-coated to provide and guarantee just the opposite.

Also, many products today are coated for manufacturing purposes before any final machining takes place. Powder coating is the process of choice for this application because of the several resin system choices available.

Type of substrate	
• Metal	• Plastic
- Steel	- SMC (Sheet molding compound)
- Aluminum	- RIM (Reactive injection molding)
- Zinc-coated	- RTM (Resin transfer molding)
- Brass	

Recommended coatings are of the thermosetting variety instead of the thermoplastic type since thermoset coatings are resistant to the heat from machining.

During the early years of my career, a manufacturer could get just about any type substrate he or she wanted to make most parts for painting as long as it was cold-rolled steel. Today, ever-increasing numbers of substrates are being used in manufacturing.

While steel is still the workhorse, several choices are available that can be influenced from the finishing perspective. For a smooth finish, a good drawing or finish-grade steel should be used. The microfinish on the substrate has a great bearing on what the eventual finish will look like.

Stamping people tend to prefer metal that is not as smooth as painting people would want. If metal is too smooth, it will not draw and form easily without splitting. Conversely, if the metal is too rough from a microfinish, the surface will telegraph through the topcoat and give an orange-peel appearance. This forces either a lot of metal finishing ahead of paint or a great deal of priming and sanding in the paint shop. This is added expense, plus it may adversely impact total emissions and other environmental concerns because additional materials are required. Work areas will extend the length of the line and require additional people, materials, and equipment, driving up costs proportionally.

There may be a need to reduce weight from a product, which would indicate consideration of aluminum or plastic. The same can be said for long-term corrosion resistance requirements mentioned earlier. These considerations reach back into the areas of stamping, welding, material handling, metal finishing, pretreatment, primers, paint, curing equipment, and time/temperature.

Many forms of zinc-clad metals are specified today. These impact in as many ways as the others mentioned. You may be asked to decide between pre-coated metals and uncoated metals that can change material handling and introduction into the paint system. For example, the spangle on galvanized versions must be considered. Types of paint products that will adhere to zinc surfaces is another variable for consideration.

More and more we find ourselves painting nonmetal substrates, such as those in the plethora of reinforced/nonreinforced plastics and composites groups. With thought up-front, the proper selection of some of these can provide for finishing them using many of the same facilities as metal. Rigid thermosetting materials, for example, can be coated with virtually equal adhesion and coverage in an existing system with mode alterations as in special systems. Flexible or thermoplastic varieties will cause changes in racking, primers, topcoats, curing time and temperature, and even paint handling equipment, if two-component materials are used.

There are many such substrates, and they each have specific needs. There can even be subsets of similar varieties. Sheet molding compound (SMC) is an example, in which various mixtures of glass content, thickness, and resin content are used. Other polyester resin systems such as those used in SMC are adapted for use in hand layup and hand sprayup of rigid parts. Each is considered a polyester-reinforced fiberglass part, and each can and does have unique qualities in the way it is produced, its tooling needs, and the quality and quantity levels normally associated with it.

Type of coating

- Liquid spray
  - Solvent
  - Waterborne
- Powder

What type of coating would make the most flexible system? Opinions in the paint community differ, but to me, flexibility should be built around liquid coating. Powder coating is here to stay, of course, and it is an excellent technology. Many improvements in powder have been made in recent years in both materials and facilities and where it can be used. But there are places where powders do not yet satisfy certain needs.

Both thermoplastic and thermosetting plastics — as well as all forms of metals — can be painted in liquid systems. Good adaptability to either water or solvent systems that meet environmental and quality requirements also argues in favor of liquid. Moreover, broad flexibility exists in the use of different cure temperature ranges for either single or plural component varieties of coatings. Powders, on the other hand, have some limitations in the low temperature ranges, and they can present problems in curing when heavy metal thicknesses are involved. Also, color changes are simpler in wet systems although powders have improved in that area recently.

These are a few of the considerations surrounding system design for liquids versus powders. Powder has some advantages, liquid has some. A thorough knowledge of which application works best for a given substrate provides the broadest state of options in system design.

In my opinion, there will never be a universal paint applicable to all types of products. The demands for world-class quality, environmental compliance, and competitive costing preclude it. There will always be niches for which specific types of coatings will be needed.

A wide variety of pretreatments is available in a paint shop as well, ranging from media blasting to simple degreasing to chemical conversion coatings. These pretreatments can be manual in nature, batch-type, or continuous-running conveyorized processes. Their chemicals may be used once or captured and used over. They may be hot or cold. They may work for one substrate only or on several varieties of substrate metals.

Type of pretreatment

- Hard wash
- Degreasing
- Chemical conversion coating

In steel manufacturing, iron phosphate is still the reliable workhorse of the industry, leading by a considerable margin over zinc phosphate and chrome conversion processes. Systems that have used hexavalent chrome are rapidly losing ground as alternative products emerge in response to environmental pressures. For quite some time, these alternative systems did not provide a similar quality level for corrosion protection, but today several are available that indicate little or no reduction in the quality levels of protection when compared to the former polluting chromes.

Zinc phosphate systems historically have offered the most flexibility for use on various metal substrates. Although they produce high quality, they do pose serious problems in the form of disposal of heavy metal sludges. The industry has been developing versions that reduce the amount of sludge, but these may be outpaced by still other technology in the form of dry-in-place products that do not contain heavy metal conversion methods now coming into use.

In sum, a host of inputs must be taken into account in selecting pretreatment facilities and products used. In addition to capital considerations, plant space availability, wastewater treatment, quality levels, and variety of substrates finished all must be considered.

By this time in the planning process, some basic issues will emerge as to what type of hanging, racking, and material transportation system will be needed. The size of parts, configurations, quantities desired, line length or plant space available, method of pretreatment, priming, finish painting (if required), and curing will determine the direction of facilities selection.

Methods of transporting the product through the paint shop vary widely in the industry. In some plants, transportation of product is 99% labor; in others, almost the total reverse. Some transportation systems are shorter indexing types, while others are long, fast, and continuous. All seem to work fine for their intended requirements.

Material handling is probably the next best contributor to success and failure in the paint shop in terms of costs. It is surprising how much "air" is being painted in many paint lines. The available window of opportunity for loading to paint areas is usually very poorly used. Many lines are under 40% full. The material handling systems are often the last to be updated, while other elements of the operations have moved forward. Product has gotten larger or changed in dimension and configuration, the substrate has changed, the paints have changed, and the line speed has been kicked upward.

When you improve the first area of waste, paint transfer and paint rework, it is necessary to include the material handling system as part of that effort. Because paint waste is the unresolved major contributor to waste and is seldom prioritized, transportation systems probably do not get even small amounts of process engineering in many places. Even though product is passing through and coming out of the line at good quality levels time after time, transfer and rework amount to 25% to 30% of excess costs.

Too many times managers are measured by yield, without having a good understanding of what is being missed in performance. Missed performance translates into cost and this extra cost has been built into the budget over a lot of years. It is not considered a rightful part of standard operating costs, yet we lose money every day as we meet our budgets, all because the premise upon which the departmental operating budget is calculated is wrong.

The first thing that must be determined in planning for paint is whether the product or chosen process requires a primer. Many products do not require a primer to satisfy quality requirements. This is true for both the lower end of quality and at the higher levels.

Additionally, many products never require corrosion protection levels for exterior exposure. Items such as tool boxes, lighting fixtures, and interior products historically have been cleaned by degreasing only and then color painted. They satisfy every customer quality requirement expected

Type of primer
• None
• Dip
• Spray
• E-coat
• Self-etching
• Autophoretic

of them. To do anything further would add to the cost and make the manufacturer less competitive.

As mentioned, shops using chlorinated degreasers will be obligated to phase them out of operation because of environmental regulations. Still, there are other alternatives coming on stream to replace such methods with aqueous cleaners that meet the new regulations. Some systems being marketed can even refit the existing degreasers to utilize the cleaners. The good news is that it appears that the capability to finish such products will be retained with very little facility modification.

Other forms of painting require no primer because of the type of technology employed. Electrocoat (e-coat) one-coat technology is one, providing a good level of performance in a primerless system. Such a system is more capital-intensive than others, but can be an excellent choice for certain niche-type paint requirements.

Autophoretic systems are used with and without primers. Certainly these are not systems that require a typical primer followed by a topcoat. Usually products are coated with only one layer of film which serves as the entire finish. This finish gives excellent corrosion resistance.

When priming is required, there are numerous methods and products available, ranging from simple dipping to more sophisticated techniques such as electrocoat. Results range from simply protective in nature to very smooth, uniform film thicknesses. Most of the systems can be arranged for either continuous flow or indexing batch approaches, depending upon volumes, space, and capital considerations. Several candidates are available in either water- or solvent-based resin systems. Powders as primers are also available in several generic resin choices.

Primers fall into a number of categories. Primer/sealers provide adhesion and color background and sealing for certain substrates, but seldom improve microfinish on a substrate. For this, a regular primer with more pigmentation or filler is needed. Sanding may or may not be required after priming, depending upon how much readout takes place from the substrate or from dirt. If a substantial amount of readout occurs, it is better to go back into the system and improve the preparation. Repeated primer operations and sanding not only increase pollution potential from the additional coatings, but usually are very disruptive through the paint shop from a constant flow point of view. Costly additional handling, incorrect materials, and increased potential for damage usually emerge when attempting to overcome serious manufacturing shortcomings through correction in the paint shop. Paint lines are for finishing, not rebuilding the product.

If a more serious need for filling does occur through sanding defects, it is normal to use a primer surfacer. These products have more pigmentation-to-resin binder to fill low spots. They sand more easily than other primers, which adds the capability to featheredge or taper into the surrounding surface.

There are also special primers such as the self-etching variety and those designed for application over flexible substrates for adhesion. The self-etching types are used when bare metal repairs have been made and the phosphate coating has been ground off. The flexible varieties are usually polyurethane and are required for adhesion to plastic parts such as automotive impact-resisting front-ends and bumpers.

Application method

- Manual/automated
  - Air spray
  - Airless spray
  - Electrostatic
  - Fluidized bed
  - Rotary atomized discs or bells
  - Air-assisted airless
  - Dip
  - Flow coat
  - HVLP

There are a number of ways to apply paints or primers either manually or automatically. However, new environmental regulations may very well reduce this number in the near future.

One of the basic mandates of the Clean Air Act is waste minimization, and not just waste treatment. I have said that the best gallon of paint is the one you do not have to use to perform the job required. That gallon of paint becomes much more important in light of the new laws.

The Environmental Protection Agency (EPA) has indicated that it will seek to have higher efficiency transfer application equipment in use in paintshops. It is the EPA's desire to see a level of 65% or more be employed. Several of the old standard varieties now fall out as potential candidates. Basically, only electrostatic and HVLP-type technologies meet new EPA requirements based on work begun in California.

Other systems such as waterborne dipping and flow coating will probably meet EPA requirements, but may or may not meet appearance requirements and other demands of the customer.

Technologies such as e-coat, autophoretic, and fluidized powder systems exceed that capacity if they are the chosen form of application. Spray methods using electrostatic or HVLP guns in either manual or automatic mode can meet or exceed this level of transfer efficiency when kept properly tuned, maintained, and used. Electrostatic rotary discs and bells will normally far exceed this requirement as well. Versions of this technology are available in guns and other application devices.

The bottom line to application methods is based on the principle of waste minimization. Getting the job done with less material must be incorporated into all future finishing considerations. By following this approach, you will make substantial strides in controlling both initial costs and back-end-loaded expenses such as maintenance and disposal.

Curing	
• Air dry	• Combined radiation/convection
• Direct or indirect convection	- Turbulators
• Infrared radiation	• Electron-beam
• Ultraviolet	• Vapor condensation

All paint that appears dry is not always cured, nor does it always provide the final quality possible from certain generic resins used for coatings. A good bit of finishing is performed using air-dry or force-dry technology. How long this will continue is becoming an issue of importance in some sectors of the finishing disciplines.

The older mainstays of the paint world — alkyds and lacquers — are being pressured out of the marketplace more and more due to the combined pressures of environmental compliance and increased market demands for quality. Some of the void left is being filled by conversion to water reducible products to meet the environmental stipulations, but demand for increased gloss retention or chemical resistance will see users moving to materials such as more fully cross-linked thermosetting resins. These resins will require the use of a proper oven matched with the paint resin chemistry.

A broad spectrum of paints and curing processes is available that permits the coating of a wide variety of substrates. This variety of substrates includes both metals and nonmetals as well as rigid and flexible products. These may be either thermoplastic (with limited heat resistance without deforming contours) or thermosetting (which can withstand temperatures up to nearly 300°F [149°C] for the non-metals and over 400°F [204°C] for the metals).

There are now means of curing films in a matter of minutes with electron beam, ultraviolet, and vapor condensation technologies. Both single and plural component products are available to meet time, temperature, and quality objectives.

In my observations over a wide range of coating industries, I have found that most older ovens are of the convection variety with some infrared units in evidence here and there. Many have been overused and abused through a combination of poor maintenance or attempts to push increased amounts of production through an old system not designed for that level of throughput: normal practice, I have noted with dismay, is to increase both the line speed and carrier loading and raise the temperature as high as possible.

This results in all forms of quality problems, from solvent pop, to discoloration, uncured paint films, loss of gloss, and orange peel. Many ovens do not have dirt shields on the conveyors; others, such as batch ovens, lose their heat each cycle when doors are opened and no time is allowed for heat-up.

It is my recommendation to check the amount of "air" you are painting when using a moving conveyor. Proper loading of a paint line will permit the slowing of conveyors to achieve much better curing, as opposed to speeding up the line.

Some operators are very good at not painting "air." Most become so accustomed to looking at certain patterns on certain hooks they do not realize how poorly they are using a system. While paint lines vary based on the size of products being painted, manually operated conveyorized lines in most cases should be designed around a 4-foot by 4-foot "window" for each painter. This area is what most painters can easily reach without moving around. Combining the line speed with the fluid delivery (volume solids) in a manner similar to the speed and feed of a lathe during machining should permit you to judge how much of the 16-square-foot area you are able to load and keep up with.

This should result in the best line speed to feed parts into an oven and maintain the proper time and temperature relationships. If you still do not have a complete cure, you may need to extend your curing oven.

To me, curing is the payoff for everything done up to that point. The quality of the finished product can suffer severely if the process either under-cures or over-cures a finish. Material handling can be heavily impacted, breakdowns will increase, and field quality will fail prematurely.

Many otherwise good products have been ruined in this way, caused in most instances by management trying to get numbers from older systems.

Many paint operations require some level of preparation either before first-time painting or to make repairs on defective products. Paint shops must be prepared and equipped to sand, mask, blow, tack, and even strip products prior to applying paint. The well run shop will obviously try to keep these operations to a minimum, as they represent additional costs and are great potential producers of dirt. There are a number of primerless systems today in both powders and liquids that help to minimize or eliminate such repair-preparation tasks. These primerless systems can produce very high quality and should not be confused with former methods that were considered to be of much lower non-glamour finishes.

Repair-preparation
• Sanding
• Masking
• Blow and tack
• Stripping

One strategy that I have found addresses the issue in a number of ways. When laying out a new system or making a major modification, you may find it advantageous to put preparation areas either side by side or even route original and repair prep through the same area.

Such type of work does not know if you are sanding something for the first time or as a part of repair. The same is true for masking, blow, and tack-off operations. Modern use of power and free conveyors will permit the movement of product on and off a conveyor and into such common operational areas. One can actually lay out separate "dead" holding areas in this fashion, even for various levels of preparation. Small amounts of work can go into an express area while more severe activities can have their individual leg of free conveyor rail. From these various dead lines, work can be appropriately rescheduled and moved back into the system.

In similar manner, many companies provide additional types of artwork such as multi-toning, striping, and decals on customer demand or as part of their own design. This is a fine potential source for adding value and increasing profitability, for both the company and the paint shop.

Once again, the equipment and system are dumb. They do not know whether you are doing this work first pass or as an added operation. It is a matter of the layout supporting the material scheduling and movement in and out of the area. If you must keep some resources available for repair and normal preparation, then why not help pay for them with value-added art operations?

Another argument for keeping such work together in one area is simply one of overview and control. This keeps supplies and equipment confined to a smaller common area, concentrates lighting, and permits easier overview and scheduling.

Always try to avoid or minimize preparation and rework as your first line of cost control and quality. In reality, there will probably always be some accommodation for these functions in a paint operation and, if so, they can be easily offset by value-added operations. A number of finishing shops actually make money in their preparation/repair areas. Different shops have different needs, of course, but such a possibility should not be overlooked.

Production rate

- Number of parts
- Number of pieces
  - Assembled
  - Unassembled
- Hours/year operation

If we are to size the throughput of any finishing operation, it is obviously necessary to have at hand complete data concerning potential product. Basic information such as how many parts, what size, whether assembled or not, and proposed hours of operation are just a few of the data bits needed by the paint manager for planning.

Through the years, the strategy for sizing paint facilities has settled into two schools of thought. In one camp are those who believe you should put in half the size you need and run it more shifts a day. On the other side are those who believe you should put in a larger operation (and capital investment) and only run it one shift a day. Each of these has merit and can be supported by valid arguments.

Let us cite an example for the first case of the smaller shop being run more hours. Plant "A" operates the assembly line or areas on a one-shift basis. The subassembly areas of the plant are also capable of producing all the required parts in one shift. This means the plant will have a storage area ahead of the paint shop to collect product being manufactured throughout the day, since it cannot paint as fast as the parts are being produced. There will also be a storage area after paint that is full each morning when the assembly line starts up. Painted parts will flow out of this storage area to the line faster than replacements flow in from the paint shop. Thus, the storage area is depleted during the day and nearly empty at day's end.

All this will work fine as long as there is no extended downtime in the paint area during the two shifts of operation. If a problem occurs in the daytime, there is a risk the assembly line will run out of product and result in significant loss of production with little flexibility to recover that day. Even if production is restored in the paint shop, product will not flow into the storage area fast enough to keep the assembly line running without stop-and-go performance.

If the assembly line were to run two shifts, there would have to be a significant reduction in line speed to match the production levels of the smaller paint shop. This would result in reducing the "banks" or storage areas ahead of and after paint, but any downtime in the paint shop would ripple through the production line all that day. It would be necessary to then run the paint shop overtime on the third shift to catch up for the next day.

This method of logic and planning is similar to walking on a razor blade every minute of every day. It can be done, but what a way to operate just to limit capital outlay. Eventually some contingency will occur and people will be pressured and mentally and verbally harangued to improve performance. This mode of operation usually results in a rapid turnover of personnel, further feeding the problem mill.

The other school will invest more money initially in a larger paint shop to avoid having these kinds of problems. Thus, they trade more capacity and capability to catch up quickly in some cases in exchange for heavier investment. These shops tend to "fudge" a bit in their "banks" or storage areas and carry 10% to 20% more inventory than required for feeding the production line. Then if any downtime occurs, they do not need to stop the assembly line. They can work some overtime on the second shift to catch up, if required.

Which one of these approaches is more correct? Neither is ideal, since both are missing a couple of considerations. It is axiomatic that business does not run on a straight and predictable line every month or season of the year. Because there are ups and downs in demand for product, there is an attendant need to provide flexibility to handle market and production variances.

It is best to size either form of facility about 20% greater than the line capacity of the assembly area. In other words, the best place to run a paint line is at 80% practical plant capacity (PPC) on any given shift. This will enable the paint shop to independently increase capacity during higher demand periods without working overtime.

Such a level further takes pressure off certain equipment by not causing it to operate at peak capacity all the time. It allows time for the paint manager to shut down and perform needed maintenance instead of running up to 20 hours per day. Moreover, this approach virtually eliminates poor performance caused by maintenance performed in catch-up mode and helps ensure that all needed maintenance and cleaning is done. (A missed cycle here, some dirty filters there, and some sludge left in the system torture performance. We live and die by doing those things that promote performance in a proper amount, on time, and on a regular basis.)

By sizing so you can run normally at 80% PPC, you get off the razor blade and have at your control the flexibility to handle the seasonal ups and downs in a more predictable manner. There is no perfect answer as regards which school to subscribe to in planning a paint facility. I have observed both forms work with varying degrees of efficiency, and those involved were satisfied. In most cases, they were comfortable with the operation and knew how to react to contingencies. However, the shops that use the 80% rule seem to be more responsive and their people much happier.

System flexibility

- Future production
- Product changes
- Sequencing
- Scheduling
- Number of colors
- Multitone capability
- Modular
- Adaptability to different generic paints

The 80% rule of PPC provides the basis for the flexibility needed in markets demanding ever faster response. But how does the paint manager determine where that 80% is? This is where leadership with long-term vision comes into play.

First of all, are we talking about today or tomorrow? What will be the production needs three years from now? Five years? Will there be a major product redesign? Is a change planned in substrates or manufacturing techniques such as moving to plastic from metals? Is there a move toward increased use of special colors, multi-toning, and different materials? What are the environmental issues to be considered?

With answers to these questions, the paint manager next must look inward.

Could you schedule your needs through the existing system(s) if any of these product requirements were to occur? Could you clean and pretreat the product if major substrate changes were introduced?

Nothing can be overlooked in your planning effort, or you will find problems lining up to happen. It is impossible in a book such as this to list types of alternatives that could be required for the myriad painting shops operating today. There are a few things, though, that could apply across the board in your pursuit of flexibility and in leveraging your capital investment.

To begin, leave enough open space in front of an iron phosphate pretreatment system for a couple of future stages. This gives you the option of an additional cleaning stage or even converting to a zinc phosphate system, if required, without tearing the entire line out and making a major rearrangement.

By making the paint booths a bit wider, you will have the space to install automatic application equipment.

If you foresee a waterborne paint system, put in stainless steel pipe and pumps.

Use separate conveyors with automatic transfers between the pretreatment and paint operations. This provides separate, independent stations for cleaning and painting and allows spacing between different operations to be changed. Carry this through to the ovens as well. You can often move product closer in an oven than required in the pretreatment or painting operations, enabling a product to continue out of the oven and not get "stuck" there if a conveyor stops somewhere else.

Make flash-off zones a bit wider in the event you want to move to waterborne products in the future. There will likely have to be heated flash-offs in that event.

Leave some space at the end of the oven to provide for expansion at some future date.

Locate the paint shop in an area on an outside wall of the building. This is good practice for a number of reasons. First, it permits expansion if major additional line(s) are ever required. Second, it helps preserve the quality of work. It is never good practice to put a paint shop in the center of an assembly operation — there are too many opportunities for contaminants to enter the paint area. It is dangerous if a fire would occur, and it could cause the loss of the entire plant. Also, this is premium space and there will seldom be room for growth. Roofs can also be an excellent place for painting shops.

In any scenario, carefully study material flow in, through, and out of any painting operation. Astute managers find ways to minimize the banks and the time materials stay in them. Use of flow charts or computer simulation can be great help in determining if any future bottlenecks might occur with any proposed change of production.

Dirt continues to generate the greatest defect rate in most paint shops, and much of the dirt is carried into the paint area on material handling devices. Next to dirt, material handling damage is the principal cause of part rejection. Though we pay attention to the paints and application, we regularly lose product and profit to these two culprits. The moral: pay close attention to material handling.

But another hazard to productivity also lurks in the wings: product positioning. In my years in the business, I have found that damage created by products positioned such that they cannot be easily painted accounts for more rework and loss of profitability than in the area of actual application.

Consistently, we have seen performance gains of 25% to 30% after examining product positioning areas in mature operations. In most cases, these areas have simply evolved slowly over the years as people attempted to preserve their original carriers or work assignments. When considered in the context of determining whether to expand an existing line or install a new one, such an examination can be doubly rewarding.

Environmental considerations		
• Air quality	• Clean room	• Respirator/uniforms
• Water quality	• Eating areas	• Grounding/safety needs
• Plural components	• Locker and shower	

Operations that have had forward-looking and responsible management have not waited until the last minute to meet the ever-increasing demands of environmental regulations.

In my observations, however, I have found the norm in the finishing discipline to be that of carping about all the negative effects and inconveniences "forced" on paint people because of some governmental ruling. It's understandable, because compliance means change and change — at least initially — is almost always resisted. And when change comes in the form of regulations, it carries an added negative of compliance.

But what these regulations should do is force the reexamination of what is currently being done and why. It is almost axiomatic that this kind of effort produces other improvements that go a long way toward offsetting any new costs associated with compliance. In many cases, such reexamination has even prompted significant cost cutting.

From a management point of view, issues in implementing change are (1) the changes are often required at a time when we would prefer not to change, and (2) the mindset that "big brother" is meddling in our business.

It seems never to fail that interruptions occur when the shop is running at the extremes of the business cycle. We are either using every resource available to produce high volumes of product and the shop just can't be modified then, or we are in a recession and money is too tight to make changes. But cash is not a valid reason for failing to meet a legislated regulation. Such a rationale is shortsighted and will result in much higher costs down the road.

The feeling of intrusion is especially great where smaller entrepreneurial ownership is involved. Having worked for many years to build a business and shepherd it through good times and bad, these entrepreneurs resent government intrusion or anyone telling them how to run their business, especially people who have no experience running a business. It can be a deeply personal feeling that literally paralyzes their action. Such inaction can lead to missed target dates, fines, or more costly last-minute fixes to meet compliance.

It is not uncommon to see some managers decide they just do not have the vigor or desire to make the required changes late in their careers. They instead choose to outsource their finishing work, which results in a loss of jobs.

In the end, it is timing and the skill and ability of an effective manager that will successfully adapt technology into the workplace. In a few isolated cases, technology may not have been created to meet the need. But in most cases, several technological alternatives are available. It is up to the manager to find the ways to integrate change with a low degree of risk to the total enterprise. Easy to say, but not so easy to do.

The "contrarian" approach

There may not be any single answer or approach to strategic planning where environmental compliance is concerned. Compliance can be achieved in a number of ways. One way I found useful may be of value to you.

There is a certain method of investment and managing today commonly called the "contrarian" approach. It calls for doing pretty much the opposite of what the majority does. When most people are buying, the contrarian is tending to sell because the price is normally high or at least rising. They know most things go in cycles, and sooner or later prices will drop, and then they once again will go bargain hunting.

Really successful people I have observed adapt this mentality to their business planning. When business is booming, they treat the customers well, sign orders, work hard at satisfying the customer, say please and thank you, and bank every dollar they can while it lasts.

People are dedicated, and work long and hard, understanding the basic rules and goals involved. They do not go crazy and spend money just because good times are upon them. When the peak is passed, they still treat the customer well, sign whatever orders there are, work hard at satisfying the customer, and say please and thank you.

The difference is they use that somewhat down period to renew themselves. They do not lay off their people. They use them to clean, change facilities, maintain the property, paint, inventory, and get ready for the next upswing. People are recognized for their efforts, given some personal time, perhaps some monetary bonus, and, very importantly, provided job security during the slow period. In simple terms, such businesses state by their actions that they intend to remain in business and are willing to create the funds during good times and then plow some back into the operation. It is wise practice, for a number of reasons.

First, it nurtures employee loyalty that will likely go unshaken because of the willingness on management's part to assure steady employment. It is during this low business cycle that a manager can obtain the best prices from suppliers and contractors. It is also the optimum time to interrupt operations to make changes. There is less risk from a customer relations standpoint and from the cash position. Less risk is involved than in doing it at high production, and you have the retained earnings in the bank to pay for the renewing.

The working environment

Environmental considerations in the well managed paint shop also include the employees' working environment. These considerations include such things as air and water quality, OSHA safety rules, and other mandated requirements.

We should also include such considerations as clean-room facilities for improved quality and employee health and comfort. Are workers provided with suitable eating areas, lockers, restrooms, and shower/wash-up facilities? Do they have the proper uniforms, gloves, respiratory protection, and relief from extremes of hot or cold weather? Are you providing for your employees the same level of environmental comfort that you afford the parts you paint?

Little things like providing something extra and cool to drink in very hot weather can mean so much to your workers. Providing an area or "oasis" that is cooling in summer or warm in winter will reap a lot of dividends. Above all, maintain a clean work place. Cleanliness fosters quality.

It is impossible in a few paragraphs to deal in depth with all the specific regulatory aspects of what is required for a paint shop. Numerous books deal with this subject in detail. If you do not have this expertise in your operation, retain a consultant or engineering firm who can provide such capability for your individual needs.

Managers must provide the climate for timely planning and actions to remove roadblocks to conformance before actual technical implementation can take place. They have to show a willingness to provide facilities and materials in a timely manner for meeting environmental regulations with no loss of product quality. Then he or she must devise the technical means of achieving this and get the plan accepted. All these must be scheduled so as to minimize interruption of ongoing operations. Finally, the work must be implemented and paid for.

The "contrarian" method described has proven to be an approach that will do these things better than other forms of operational management. This method is especially beneficial to midsize and smaller companies whose ownership and control is vested in a small group of individuals as compared to absentee ownership or management not directly involved with close day-to-day operations.

Any failure to meet environmental regulatory requirements in the paint shop lies on the shoulders of upper management. It is the responsibility of management to provide the funds after a solid technical and value-related program has been developed from the several alternatives available today.

The bottom line is quite simplistic. Despite all the technical competence and money a company might have, if it does not bring itself into environmental compliance properly and on time, it simply will not obtain a permit to operate in the future.

The former conventional wisdom that environmental dollars are nonproductive dollars has to be avoided. You, as a manager, should be upbeat and positive. The alternative is not appealing. We are seeing companies closing down and management actually going to jail in the more blatant cases of environmental laxity.

Responsible managers do not abuse the environment for the sake of a few dollars more profit in the short term. By taking their responsibility seriously, they ensure the long-term viability of their organization and the communities around them.

Paint delivery systems

- Pressure pots
- Circulating systems
  - Tanks
  - Tote tanks
- Material transportation

Any paint shop will need some means of delivering the coating materials into and around the application area(s). These could include items such as spark resistant lift trucks, carts, dollies, and even automatic transport vehicles (ATVs). It could even be as simple as a worker carrying a bucket to a small paint pump. But whatever the size of the operation, there has to be some means of transporting materials.

Since numerous books and many firms are available that detail material transport, there is no need to design a paint delivery system here. Our focus instead will be on the role of management in determining specific needs.

Management must select from a wide range of systems for both liquid and powder technologies. Determination must be made between solvent- and water-based liquid systems. Within the system itself are alternatives ranging from simple siphon cups to traditional pressure pots. Circulating systems can be tailored to specific needs from small "junior" units to major ones operating from buckets or drums or tote tanks. Piping may be conventional or stainless steel, depending upon the generic resin systems employed.

There would normally be a requirement for an approved paint storage and mixing area (room). Mixers and other equipment (such as lights) will have to be explosion proof or able to operate on compressed air to satisfy most safety and insurance considerations. All product and material conveyance equipment must be grounded. Good lighting should be provided.

Provision must also be made for support elements such as regulators, flowmeters, agitators, and filtration as part of any facility consideration. Both environmental and material handling considerations might indicate a need to recover solvents with the use of a still.

Other issues: Can you schedule materials in a JIT mode to reduce storage space needs and subsequent waste storage requirements? Do you need paint heaters to either reduce VOCs or provide control for higher solids-compliant coatings?

Once again, managers must be aware of their needs and then ensure that they are provided for and well maintained. They should stay alert for improvements in both materials and facilities that are well mated and user friendly. Frustration levels rise and productivity falls when parts are difficult to obtain and equipment is not reliable.

The good manager will make the effort to ensure that the source of supply is stable, technical assistance is good, suitable replacement parts are available, and these parts last for an acceptable period of time. Some equipment as engineered requires major replacement of parts. The initial price may be lower, but the long-term cost is higher and a lot of downtime can result.

The most sophisticated booths and application equipment will not perform when paired with a supply system that cannot get the proper material to them on time, in the proper amounts, with the proper solids, paint temperature, and viscosity.

This is like having a good heart and bad arteries, or even the other way around. You need both to be a healthy and performing entity.

Certainly, any manager has to provide facilities to make the paint shop function; he or she also must provide various sources of energy to drive them. We all use energy, and although it is second nature to us, we should not be so casual about it.

One of the real edges the U.S. has over other countries in the world economy is relatively inexpensive energy. But where formerly we used energy wantonly, there is now a better

Energy sources/availability

- Air
- Water
- Gas/propane
- Fuel oil
- Steam

understanding by increasing numbers of managers that energy needs to be carefully managed.

In most areas of the country, there is no real shortage of any energy source. There are areas where one form may be preferred or more abundant than another; but seldom is it a matter of not being available.

When the energy crunches of the 1970s hit industry, few plants were prepared for the lack of energy. Plants were often limited as to how many hours they could operate. Some were shut down for periods and allowed only enough energy to keep them from freezing or otherwise causing damage to the facilities. No production was permitted.

This has caused managers of today to plan for and provide backup forms of energy in many cases. We now see propane farms behind plants as backup to natural gas. Some companies have invested in co-generation of electricity.

In a much broader manner, the painting industry has continually sought products that could be used at lower energy requirements. These include sealers, pretreatment and cleaning products, paint booths, ovens, primers, and topcoats.

Paint operations today that are producing the same square footage of product as several years ago should be using considerably less energy, thanks to improvements in materials, methods, and machinery. The effective manager must continually stay abreast of energy-saving technologies because the cost of running a shop will likely not go down.

No one knows what government is going to do about new taxes on energy. It is difficult to imagine that it will continue to overlook this huge source of potential revenue in the future in light of its increased use. Too many examples from other countries show how energy policies can result in much higher taxes than ours.

This situation is again exemplary of the frugal manager: the energy not used is the best energy, just like the best gallon of paint is one not used to get the job done. Common sense tells incontrovertibly that efforts to reduce energy consumption are both a good short-term and long-term policy, because costs are going up.

CHAPTER 11

Monitoring for Cost Control

Waste is a part of nearly everything that occurs. In the business arena, constant vigilance to uncover it and unrelenting efforts to minimize it are essential to maintaining competitive position and environmental compliance.

Managers today are faced with almost never-ending changes in both everyday operations and long-range plans. To gain and maintain control in an operation in such a climate is a continuous challenge. Indeed, control itself varies with each operation. But whatever the definition, control must be practiced as a dynamic rather than a static concept. And that concept must be one that primarily includes people rather than things.

While plans can be made at the executive level *ad infinitum*, not the slightest amount of actual control can be exercised without superior executive and management talent. A plethora of tools is available to help gather information in formalized methods which serve as aids and guides to management. But, in the end, the very nature of a competitive industry is such that no formalized system has ever been devised that ensures and replaces superior knowledge to make correct and timely business decisions.

In a well run finishing operation, certain fundamental controls must be recognized from both an operational and accounting perspective. There is need for a technical person to perform and monitor process engineering activities on all shifts. Material should be "pulled" through the facility, and only that which the system uses should be replaced.

Process flow and cost drivers

Process layouts should show material flowing in the direction of the shipping area, and decisions should be driven to the floor level for painting and other schedules. An informed manager knows the cost drivers and their importance to the operation. These drivers derive from data collected; therefore accurate and comprehensive records of all work performed, including rework, must be kept. A schedule of regular preventive maintenance must be established, along with methods of accomplishing it and a record of time and place provided. From these records the manager can formulate the predictive maintenance plan for the shop, which, if used correctly, can provide substantial cost and schedule benefits. Lastly, and importantly, health and safety monitoring of both the workplace and personnel is vital to both performance and morale.

The cost-efficiency ratio

These formal systems need to be in place to assist good management in dealing with two of the major elements affecting productivity in the paint shop. These two elements are controlling costs, including the elimination of waste, and increasing efficiency as part of waste minimization and meeting environmental regulations.

Every well managed painting operation focuses on these two activities. Though slight differences exist shop-to-shop, the common thread is that well versed managers understand their importance as cost drivers and their criticality to meeting product numbers, quality, and environmental regulations.

In all human endeavors, some kind of waste is generated. While it is normally not one of the more popular topics of conversation, it is an item that must be dealt with. Everyone is responsible for providing a means to handle the problems of waste.

What we should try to do is minimize the waste and associated problems it causes, such as disease, inefficiency, and eventual depletion of resources. In the paint shop, waste is a perpetual enemy. We generate a lot of waste even when we operate under controlled conditions and are painting "efficiently." Waste reduction is sensible and economically rewarding in any business, but in an era of environmental sensitivity, it becomes a priority.

As these mandates become even more restrictive, we as finishers will have to make changes in what we do, how we do it, and what materials we do it with. It serves as an opportunity to reexamine our entire operation for potential improvements.

Variables impacting costs

We want to monitor, correct, and install new guidelines for operations. These become necessary as facilities and materials are married. So let us focus on the variables in paint operations to obtain the best "total costs" for an operation as compared to looking at everything on an individual basis of cost. Gaining insight into best total cost is a time-proven practice of good management. If more managers practiced this concept of limiting waste and observing total costs policies, there would not be so great a problem of waste management today.

When we discuss efficiency, it is not just transfer efficiency that we are looking toward. Efficiency encompasses all the variables that directly affect the quality and productivity of the operation. Without the understanding of all these variables, it is impossible to obtain the greatest degree of efficiency and thus reduce costs and waste.

Figure 11-1 lists most of the major variables found in a modern paint operation.

These are the elements the paint shop manager must deal with. It is impossible to run an efficient operation on a daily basis without taking these elements into account. We are all familiar with these variables, so a detailed examination of each is not necessary here. The importance of the list is the number of items it contains and the considerations that must be made to control the ultimate performance, quality, and cost of finishing a product.

Some of these more than others tend to be ignored in some paint shops, so we will highlight these as the beginning considerations for improvement in performance.

Most poor finishing operations tend to paint too much air in relation to product. In these shops, the product is either poorly hung for coverage, not as densely located as possible, or improperly spaced carrier to carrier. It is not uncommon to find as little as 25% of the correct capacity being utilized. Often, two and even three times the capacity and greatly improved efficiency can be achieved through better hanging and material flow practices.

It is often possible to greatly reduce the line speed and obtain better flash-off and curing besides obtaining better transfer efficiency during application. Sometimes it is possible to reduce the number of shifts and even postpone modifications that were indicated before instituting loading changes.

A number of years ago while I was working on a project in Europe, such a situation emerged. The plant needed to produce 52 vehicles per day in a two-

Finishing System Variables

- Conveyor line speed
- Conveyor loading efficiency
- Racking
- Pretreatment:
    - Washer design
    - Washer maintenance
        - Mechanical problems
        - Chemical problems
        - Quality problems
- Water entrapment
- Air blowoff station
- Dryoff oven
- Cool-down zone
- Temperature of ware entering spray booth(s)
- Spray booth type
- Spray booth size — air movement — volume, velocity
- Climatic conditions
- Temperature of coating
- Viscosity of coating
- Coating rheology
- Solvent system
- Operator knowledge and techniques
- Application equipment
- Fluid flow rate/gun — fluid pressure
- Atomizing air pressure
- Turbine RPM's disc
- Disc flow rate — oz/min
- Reciprocator speed
- Vertical triggering efficiency
- Horizontal triggering efficiency
- Manual touchup operator efficiency
- Fan pattern size
- Distance application device to ware
- Paint transfer efficiency basic
- Wet paint film thickness
- Dry paint film thickness
- Color change efficiency
- Fluid handling system
- Flash zone
- Bake oven — time, temperature, mass efficiency
- Cool down zone
- Make-up air distribution
- Reject level, dirt and other defects
- System maintenance
- Personnel training
- Coatings, value analysis
- Paint film rejects, causes and cures

Figure 11-1. Consistent high quality finishes start with good paint line control, and constant fine-tuning is the key to quality, productivity, enhanced profits, and environmental compliance.

shift operation, but they were unable to produce more than 45 units with their current facilities. As such, a request for approximately $5,000,000 U.S. had been made for capital facility modifications to expand the paint capacity.

I made a detailed review of this operation in the first two days after my arrival. What I found was this. The conveyor speeds had all been moved faster and faster over the years to obtain more volume. They were now at a point where the times were not proper for operation of the pretreatment line stages and application of paint as presented, and oven curing time/temperature relationships were not in proper relation.

To their management's great surprise, I first requested that the conveyor speed to be reduced approximately 25% from the present level so these other variables could be brought back into control. I then announced, again to their surprise, that there would be no need for the capital funds either. Their question was, how was a solution going to come out of this rationale?

I told them that we were going to completely change the loading patterns and scheduling through the system. This generated further disbelief. How could people handle and load that much weight onto fewer carriers, paint the product, and keep up, and then unload such numbers of products after painting?

I explained. Currently, many parts were being hung as individual parts for each unit and all were the same color. (Bumpers were a good example.) In the future, I proposed that three bumpers be hung on a carrier instead of one. In this manner, we could paint three units worth of production on a single carrier and gain two hooks.

We changed the direction the parts were hung from crossways to the conveyor to parallel to it. Instead of sprayers painting between individual parts, they were now positioned in alternating stations horizontal to the parts moving by them. Each painter was then responsible for only one side of the more heavily loaded rack facing them. This spread painters out so they did not have to get paint on each other and created proper flash times between coats right in the paint booth.

This type of revision was applied to a number of small parts common to all vehicles. In the end, it was possible to work at 25% less conveyor speed and achieve a new plant capacity of 64 vehicles without any major capital outlay or facility changes. I recommended that about $50,000 worth of new hooks, pretreatment washer plumbing, and other maintenance improvements be installed.

Along the way, this new scheduling and loading approach also eliminated approximately 15 people on each shift in what appeared to be an extra benefit to performance. Instead of spending huge amounts of capital, plant capacity had been greatly increased, and there had actually been this nice improvement in performance as well. This left me with a really good feeling, initially, but some sadness in the thought that 30 people might lose their jobs.

If this had occurred in a typical U.S. operation, the floor people would have been quite upset with an outsider coming in and making revisions that might cost them their jobs. Before leaving, I wanted to talk to the people who were going to be affected and explain how and why this was taking place.

Instead of finding these people angry about their supposed misfortune, they were happy, and each thanked me for what had happened. Now it was time for me to be surprised. They were elated that their company would prosper, even if it was at their personal loss.

As it turned out, in this particular country it was practically impossible for a company to lay off anyone in this manner. There were numerous rules of job protection. Although these people were now considered redundant, they still had to be paid for a long period of time while it was seen if normal attrition would take them off this status. If this did not happen, then the company had to seek permission from the government to lay them off and hearings had to be held. If it was then decided these jobs no longer existed, there were still further benefits due the employees.

This scenario has been repeated in other places in the world under governments different from that in the U.S. My first thoughts were these forms of management theory were not competitive, their costs were out of line, and eventually such practices will force bankruptcy. They could not compete with our cost-control efforts or productivity.

These thoughts were borne out as one by one the countries have changed their philosophies. Some are ahead of others in their efforts to become more capitalistic, and hence more efficient.

Profitability or ideology?

There are many societies that believe industry has more reason to exist than just to make money, that it has certain social responsibilities, such as providing employment, even if profits are lowered by so doing. But how does a manager, faced with the situations described in the example, go about lowering costs and competing?

One way I have observed is to continue to make a product with the improvements implemented. In this way, additional product could be produced in still higher schedules to put the redundant personnel to work. The extra product made would drive down unit cost further.

This extra product was seldom ever sold within the markets of its native country but, predictably, was sold, or even dumped, in other countries. Often the exported product was sold cheaper than at home. Sometimes major producers in the importing countries are impacted even to the extent of losing their domestic market.

All this may not mean much to anyone until it becomes his or her ox that is being gored. But it does bring into focus the fact that we are not the only ones in the world seeking ways to improve productivity. A simple thing such as not painting as much air through a system can create new capacities. New capacities can be the seed for improvement that can upset an entire industry or market somewhere else in the world. When it happens to us, it can get very personal.

Application systems and transfer efficiency

All finishing people should realize that we usually do not paint metal. We prime and paint over pretreatment. Most products will carry some form of conversion coating from their pretreatment systems. Much of the corrosion protection, adhesion, and application success comes from such coatings. Large amounts of energy are consumed in these processes, as are chemicals whose disposal has to be dealt with. Without control of this operation, world-class quality and cost efficiency will be elusive.

The control and use of organic coatings is certainly of major importance in the paint shop. It is that moment of truth when all the value-added operations have been completed. It is something akin to a dog chasing a car. When he has finally caught the car, he has a mouth full of car. But what does he do with it? In the paint shop, product is coming to the application booth. All the variables of the material must be coupled with the application equipment variables. Whether it is performed manually or by automatic equipment, every effort must be focused on applying the coating with the highest transfer efficiency possible. Today, such levels of efficiency usually mean certain types of systems.

Systems such as powder, e-coat, autophoretic, dip, electrostatic, and HVLP all provide 65% or better transfer efficiency when properly controlled. The range of technology is wide enough so that most paint and finishing managers are able to select a system that is both affordable and suited to their specific needs. Studies reveal that these systems normally pay for themselves fairly rapidly.

Incremental improvement

It is quite possible that paint shop operators will have a difficult time obtaining permits to operate in the future without one or more of these technologies. Again, it is important that managers become knowledgeable about these technologies and look to the great potential of making incremental continuous improvements.

One way to move toward continuous improvement is to add a technical person responsible for monitoring and managing the variables. Next is the establishment of a small team consisting of a cross section of production, quality, scheduling, and maintenance personnel. This team should contain both management and floor workers and have a champion as the chairperson. The technical leader would be a good choice for chairperson.

This team should conduct a complete and systematic analysis of the entire paint operation's activities. If we use a pretreatment system as an example, we find that this analysis begins to reveal the weak spots in the systems and creates candidates for improvement. From this analysis, a troubleshooting checklist can be compiled (Figure 11-2). Remember, failure of the weakest item causes the whole to fail. If the team does not have the technical expertise, a consultant can be retained to do the initial review with the team, help establish priorities and schedule action items. As the team learns from this training and exposure, members can take over the activities from the consultant.

Too many times people tend to overreact or try to make some mighty problem out of correcting and improving paint variables and expect a white knight to come to their rescue. Seldom is the problem of the magnitude needing such a solution. Rather, there is more of a need to understand the problem thoroughly and do a series of small simpler acts regularly and on time. Much of the success in controlling paint area variables centers on daily effort by a well trained and dedicated work force, including management.

Regular attentive and incremental improvement will prevent the claps of thunder and bolts of lightning. There will instead be a quiet and steady evolution toward measurable improvement. When you look back whence you came, you will be surprised at how your course has changed for the better.

Monitoring for Cost Control

PROBLEMS	Temperature low	Cleaner solution spent	Spray pattern misaligned	Pressure too low	Problem drawing compound	Long drain time	Parts caught betw. stages	Low activator	Direct fire burners	Excessive dragout	Drippage	Short contact time	Cleaner concentration low	Improper hanging	Overflow reduced	Spray pattern misaligned	Improper water feed	Drag in	Drippage	Drain time	Improper hanging
Overall blush corrosion	x	x	x	x			x	x			x	x				x				x	
Top edge vapor corr.				x																	
Low phosphate coating wt.	x	x	x	x	x	x	x		x			x	x			x					
Blue ctg. appearance					x																
Dusty coating						x		x		x					x	x	x	x			
Streaks in coating	x	x	x	x	x	x	x			x	x	x	x						x	x	x
Large, sparse crystals					x		x			x											
Water break pattern in ctg	x	x	x	x	x				x		x	x	x			x		x		x	
Nonuniform coating	x	x	x	x	x	x	x	x	x		x	x	x			x	x	x	x	x	
Uncoated areas	x	x	x	x	x			x	x		x	x	x			x			x	x	
Free acid too high																					
Free acid too low								x							x	x	x	x			
Vapor corr. bel. holes, etc.	x	x	x	x	x	x	x	x			x	x				x				x	
Yellow stains					x	x	x	x												x	
Excessive sludge	x		x			x			x												
Excessive scale						x															
Excessive phos. cons.															x		x	x			
Excessive accel. cons.																					
Poor salt spray	x	x	x	x	x		x	x	x		x	x	x			x	x	x	x	x	
Poor humidity	x	x	x	x	x		x	x	x		x	x	x			x	x	x	x	x	
Poor physical adhesion	x	x			x		x				x	x				x	x	x	x	x	
Blistered paint after cure		x		x		x	x		x		x					x	x		x		
Stained paint after cure									x											x	
Chrome spot-rept. work																					

Figure 11-2. In-depth analysis of the entire paint operation will reveal system weaknesses.

PROBLEMS	PHOSPHATE STAGE	Concentration low	Spray pattern misaligned	Temperature low	Temperature high	Low accelerator	High accelerator	Leaks or overflow	Dragout	Drag in (alkaline)	High spray pressure	Chem. add. too close together	Drain time excessive	Drippage	Free acid high	Contact time short	Chemical control	Metering pump problems	Free acid low	Improper hanging
Overall blush corrosion		x		x		x							x		x	x	x	x		
Top edge vapor corr.		x	x		x	x										x	x	x			
Low phosphate coating wt.		x	x	x		x	x	x	x		x					x	x	x			
Blue ctg. appearance		x	x	x		x				x						x	x	x	x		
Dusty coating			x		x	x			x		x					x	x	x			
Streaks in coating			x			x				x			x								x
Large, sparse crystals		x		x		x										x	x				
Water break pattern in ctg		x	x	x						x											
Nonuniform coating		x	x	x	x	x				x	x					x	x	x		x	
Uncoated areas		x	x	x		x				x				x			x	x			
Free acid too high					x		x	x								x	x				
Free acid too low		x					x			x						x	x				
Vapor corr. bel. holes, etc.		x	x		x					x			x		x	x					
Yellow stains		x	x		x		x	x					x		x		x				
Excessive sludge				x		x			x	x						x	x	x			
Excessive scale				x							x					x	x				
Excessive phos. cons.			x		x	x	x	x	x	x	x	x		x			x	x	x		
Excessive accel. cons.			x			x	x			x							x	x			
Poor salt spray		x	x	x		x	x	x		x	x	x			x	x	x	x	x	x	
Poor humidity		x	x	x		x	x	x		x	x	x			x	x	x	x	x	x	
Poor physical adhesion			x			x		x			x				x	x	x				
Blistered paint after cure		x	x	x	x	x	x			x	x	x	x	x	x	x			x	x	x
Stained paint after cure																					
Chrome spot-rept. work																					

Figure 11-2. (Continued.)

Monitoring for Cost Control

PROBLEMS	PROCESS/Causes	PHOSPHATE RINSE STAGE	Overflow reduced	Water feed low	Spray pattern misaligned	Drag in	Drippage	Drain time excessive	Improper hanging	FINAL RINSE STAGE	Concentration low	Concentration high	Spray pattern misaligned	Drippage	Chrome effected	Improper hanging	POST WATER RINSE	Insufficient spray	Overflow low	Spray pattern
Overall blush corrosion			X		X			X							X			X		
Top edge vapor corr.				X																
Low phosphate coating wt.																				
Blue ctg. appearance																				
Dusty coating																				
Streaks in coating							X									X				
Large, sparse crystals																				
Water break pattern in ctg																				
Nonuniform coating																				
Uncoated areas																				
Free acid too high			X	X		X														
Free acid too low																				
Vapor corr. bel. holes, etc.			X	X	X	X		X												
Yellow stains							X					X	X	X	X			X	X	X
Excessive sludge															X					
Excessive scale																				
Excessive phos. cons.																				
Excessive accel. cons.																				
Poor salt spray			X	X	X	X	X	X			X		X		X					
Poor humidity			X	X	X	X	X	X			X	X	X	X	X			X	X	X
Poor physical adhesion											X	X	X	X	X					
Blistered paint after cure			X	X	X		X					X	X	X	X					
Stained paint after cure							X	X				X		X						
Chrome spot-rept. work											X	X	X	X	X					

Figure 11-2. (Continued.)

Summary

Variables are the bricks that must be bound together by the mortar of human knowledge, skill, and determination to excel. Without that mortar, your edifice will be shaky indeed. Control starts with each of us. It comes from people skilled in the use of systems that provide the best possible quality of information, or "QOI."

A QOI condition, when combined with desire to respond, should provide in-depth, accurate knowledge of all the variables. This then permits the prioritization needed for response and control of costs.

Certainly, controls are essential in any well managed operation. Once operating procedures have been established, it should be the goal of operational personnel to both operate at the expected level of control and to constantly assess how costs can be reduced without penalty to product quality or safety.

Operating within the constraints of paint shop variables is crucial to cost-effective painting. The many variables that affect paint operations must be accorded a high level of attention by the manager. Some, of course, will always carry higher priorities than others, but all are important. The well managed operation functions in harmony with these variables when the manager understands the importance of cost drivers in meeting goals.

CHAPTER 12

How Good is Good? ISO 9000, SPC, and Quality Control Performance

A disturbing product of American manufacturing's lack of performance is a breaking of the implied contract between its managers and their workers.

For the first 25 years of my career, there was a general feeling in American industry that you couldn't have high quality *and* high production. I did not subscribe to this belief, and it was a constant struggle to turn it around. In fact, if you believed in producing a quality paint job, you were usually considered some sort of enemy to production. But this didn't stop some people from continuing to prove that such beliefs were wrong. Good quality enhances production, not restricts it.

Though improvement after improvement was made over the years, the inroads made were never fully appreciated by many. There are times when an agent of change is not popularly received. It is normal for about 50% to swear by your efforts and about 50% to swear at you. As long as it stayed about that ratio, something good was probably happening. We were making improvements in quality, and with that effort came improved cost performance.

Is it really necessary to have this either/or situation? I think not. It took a long time for American industry to wake up in the 1980s to realize that it was losing the competitive war to other parts of the world.

Most quality systems in the United States are built around one of two basic theories. They are absolutely different from each other in how they achieve quality goods, yet both can lead to greatly improved results. Each requires more than just the application of certain techniques and gathering and analysis of statistics.

The rise and fall of Taylorism

Most American business has been driven by the ideas of a great early innovator named Frederick Taylor whose theories of scientific management were developed near the end of the last century. The effects of his theories have been felt in our workplaces ever since, and many present-day managers are not even aware that the management philosophies passed down to them can be traced to Taylor. The use of human or electronic surveillance of workers, spreadsheet analysis, time and motion studies, and ergonomics all came from the thinking of Taylor.

Taylor's theories achieved dramatic results in their day, but they contain at least two major flaws which make them less viable today. One is the disregard of the importance of individual workers, and the other the heavy dependence on management time and skill in directing every movement of the workers under their supervision.

Some companies have recently moved to having computers or other forms of electronic surveillance, such as cameras, to report and measure workers' use of time and activity. This has led to legal action and ethical concerns, along with unnecessary stress and lowered productivity on the part of the worker. Joseph Juran was an expert in the Taylor tradition and tried to humanize Taylor's methods to some degree. Still, he admitted that any program based on these methods has to be almost perfect or the process can run at a high level of on-going waste.

The other problem with Taylor's methods is a failure to take into account the inevitable differences between people, circumstances, and contingencies that almost certainly will occur. Any of these can skew calculations and forecasts to the point that they become worthless as a basis for decision making. Recognition of this flaw led to formulation of other philosophies.

Enter Dr. Deming

American industry began hearing about a man named Dr. W. Edwards Deming and what he had done in Japan. Here was a man who had used statistical quality control (SQC) quite successfully in U.S. war production during World War II and worked in Japan after the war as part of a census team. American industry did not pick up on his methods, but the Japanese invited him to teach SQC there. The Japanese Union of Scientists and Engineers recognized the importance of this technique to rebuild their country.

In contrast to Taylor's methods, Deming's approach emphasizes the overriding goal of reducing complexity. Axiomatic in this approach is that unnecessary complexity is found at all levels of a company's operation, and it can always be reduced. The results move from improvement, to good, to often spectacular.

Deming feels that endless long-term planning is a pointless exercise because it is impossible to know all the factors that will affect any complex mix of figures. While planning is important, this activity must have some limitations and flexibility.

Deming sees long-range planning as the art of adjusting to change rather than precluding it. Constant adjustment is a strategy for success.

What is interesting in looking at the two management methods is how the U.S. and Japanese reacted to each. The Japanese had always had a national tendency to be very controlling and paternalistic in relation to their workers. Yet they accepted the Deming philosophies that people were important and flexible enough to make such continual changes.

The United States has always believed in the individual's ability and right to make his or her own mark in their work. Yet we did not carry our beliefs into action. We continued to embrace the Taylor methods, which are restrictive and fail to make use of the individual's potential as a problem solver and contributor.

Deming's premise was that productivity increases are directly proportional to quality improvement: defective product is waste, pure and simple. It is a simple premise, but a profound one. There is less rework and waste when work is performed in a quality-minded operation.

Not only does quality improve the balance sheet, the improved productivity can lead to obtaining more business because of potentially lower costs. It also provides workers with a sense of pride and accomplishment in their own abilities and their future.

American management's failure to recognize the validity of Deming's theories has led companies to break the implied contract between them and their workers, a contract understood to mean that the company would provide certain forms of security for them and the community in exchange for loyalty and work. Instead of getting better through continuous improvement, many companies opted to pull out of communities and go to other areas or countries chasing cheaper labor to lower costs. They refused to change their style.

It was refreshing when suddenly there was a champion expounding security and continuous improvement who was armed with the performance record and visibility to catch executive management's attention.

Measuring quality

When I first started working in paint 40 years ago, more than 50% of everything painted had to be reworked to some degree. This was the accepted standard, and much attention was given to how rapidly repairs could be made, instead of eliminating them in the first place.

The idea of eliminating errors was a hard sell for many reasons. Mainly, it was quite difficult for the accounting systems to pick up, document, and fund this effort from a justification perspective. It was a lot easier to document a reduction of time and materials in the actual repair activity than to do so for something that never happened in the first place. Many is the time I felt like a salmon swimming upstream.

Today, in contrast, it is most pleasing to work in operations that perform on a very routine basis of 2% rejection, and efforts are still being made to improve that every day. This is a tremendous change!

On the subject of quality, there are available to the manager today many tools and directions that produce results. If they are not currently a part of your daily business, it is safe to say some will be in the very near future.

Quality is measurable, but how good is good? It varies from product to product and how the product is to be used as well as numerous other considerations determined by the marketplace. But there is some common ground for measurement. There will always be a leader in the world for a particular product or niche in the marketplace. This world-class leader — or benchmark — will cause the others competing in that market to assess their product offering against the leader.

More and more, the level of quality is not the minimum of a specification. It is a dynamic level that is changing based upon the benchmark products. It is a moving target, always higher than what had been considered the specification. This in real life has become the measure of "how good is good" in more and more cases.

ISO 9000

Any discussion of quality and competitiveness today would have to include the ISO 9000 quality system standard. The worldwide growth of this program is phenomenal, yet much confusion surrounds it and the implications it may have for us.

This chapter is not intended to teach anyone an entire course in ISO 9000. Instead the intent is to provide a condensed outline and description of what it is and what it is not. Since numerous books and courses are available for specialists and users alike on this subject, it is my recommendation that paint managers obtain this level of familiarization of ISO 9000 from these sources. Many of you are going to have to be a part of registering to these standards in your paint operations as part of an overall plant effort.

ISO 9000 is a series of quality management and assurance standards. It was developed by a technical committee under the International Organization for Standardization, or ISO. The standards were first published in 1987, with the U.S. version called ANSI-ASQC Q90. (The initials represent the American National Standards Institute and the American Society for Quality Control.)

The ISO 9000 series is broken down into five subsections:

- ISO 9000 (ANSI/ASQC Q90) Quality Management and Quality Assurance Standards - Guidelines for Selection and Use.
- ISO 9001 (ANSI/ASQC Q91) Quality Systems - Model for Quality Assurance in Design/Development, Production, Installation, and Servicing.
- ISO 9002 (ANSI/ASQC Q92) Quality Systems - Model for Quality Assurance in Production and Installation.
- ISO 9003 (ANSI/ASQC Q93) Quality Systems - Model for Quality Assurance in Final Inspection and Test.
- ISO 9004 (ANSI/ASQC Q94) Quality Management and Quality System Elements - Guidelines.

The system can be used for either manufacturing or service organizations as a means of applying standards to managing and improving operations.

The initial interest was driven by demands from the European Community (EC) to U.S. companies who wished to do business there. Many European companies prefer suppliers to use this as their quality standard.

The ISO 9000 method has now spread beyond the original EC, and is becoming global at a rapidly increasing pace. Pressure is also coming internally in the U.S. from customers hoping to assure top-quality products.

ISO 9000 standards are unlike normal engineering standards like test methods, measurement, or product specifications we have experienced. The concept is that certain generic characteristics of management practices could be meaningfully standardized. This would then give mutual benefit to producers and customers alike. ISO 9000 standards are tools to achieve total quality management.

More than 50 countries have adopted the standards, and 32 more currently have third-party assessment and registration services. The waiting list to obtain qualified approval sources is long, resulting in similarly long delays in some countries.

Surprisingly, the set of documents is quite short. Given the broad reach of the standards, one would be led to believe this was some huge document written in typical bureaucratic garble. It isn't. ISO 9004 is the longest of the subsections, with 16 pages of text. ISO 9001 and ISO 9000 contain seven pages each, and ISO 9003 has two pages.

The ISO 9004 gives guidelines for developing and implementing the requirements of ISO 9001, 9002, and 9003. It also includes considerations on economics of quality in several disciplines within an organization.

Products are identified which are considered regulated or unregulated. Fourteen of 23 categories are intended to be regulated by the Europeans. Some products are considered mandatory such as health and safety items. Eventually no product produced anywhere in the world can be sold in the EC without the "CE" mark of conformity.

Registrars for suppliers currently must be accredited by a European accreditation body. U.S. companies that produce unregulated products will still be expected to comply with ISO 9000 in two more years.

The critical question looming in the minds of manufacturers around the globe is what are the costs associated with ISO 9000 registration. Product liability, plant registration, and third-party assessment of products are some areas, to mention a few. Many feel there are lower unit cost potentials to be gained from the improved quality that will more than offset these costs. Each must determine this on an individual basis. Many small U.S. manufacturers are concerned they will get blocked out of the EC market by not being able to afford ISO 9000 compliance.

There is a difference between "quality system registration" and "product certification." Quality system registration applies to a firm's quality system as a whole and is voluntary and mainly market driven. Product certification focuses on certain products and is regulatory and must be performed by "notified bodies," a type of third-party certifier in the EC.

Self-Certification or Self-Declaration is common both in the U.S. and Europe. It certifies that the supplier assures the reputation of its products' quality and delivers it.

Third-Party Registration is a means of demonstrating product quality by using an independent auditor to conduct tests. The "CE" marks are different for each method of certification.

When considering selection of a registrar, be sure its certificate will be acceptable to both customers and regulators. Ensure that the registrar be able to deliver a certificate on time and that they are stable financially. If they go out of business, it will still be necessary to go back every six months to determine if you are still qualified.

Importantly, make certain that they are experienced and knowledgeable about your product and processes. In general, ISO 9000 standards are scheduled to be reviewed and revised, if necessary, every five years.

The Registrar Accreditation Board (RAB), an affiliate of the American Society for Quality Control (ASQC), accredits organizations to become quality systems registrars in the U.S. It is important to note that they certify organizations as registrars, not individuals.

ISO 9000 is a customer-driven effort for continuous improvement. In this, it begins to take on some of the Deming approach. While Deming's steps are geared to statistical orientation, they also include management concepts in the form of his "14 Points of Management Obligations."

Management's obligation

Dr. Deming, in his book, *Out of the Crisis*, states that "management has failed in its efforts to plan for the future and foresee problems." This has brought about great waste, loss of markets, and unemployment. It is his further feeling that all the activities classically performed by management — such as problem solving, increased automation, expansion, increased use of computers and other gadgets, and the explosion of Q-Circles — will not halt the decline in American industry. Only transformation of the American style of management and government regulations can halt the decline.

It is in this push for transformation that we should look at his "14 Points of Management Obligations" (see Figure 12-1). These can apply to organizations large and small as well as to a paint shop within a company. The first of the 14 points defines what has been my feeling of how all in business should perform.

Create constancy of purpose toward improvement of product and service, with the aim to become competitive and to stay in business, and to provide jobs.

131

14 Points of Management Obligations

1. Create constancy of purpose toward improvement of product and service, with the aim to become competitive, stay in business, and provide jobs.
2. Adopt the new philosophy. We are in a new economic age. Western management must awaken to the challenge, must learn their responsibilities, and take on leadership for change.
3. Cease dependence on inspection to achieve quality. Eliminate the need for inspection on a mass basis by building quality into the product in the first place.
4. End the practice of awarding business on the basis of price tag. Instead, minimize total cost. Move toward a single supplier for any one item, on a long-term relationship of loyalty and trust.
5. Improve constantly and forever the system of production and service to improve quality and productivity, and thus constantly decrease costs.
6. Institute training on the job.
7. Institute leadership. The aim of supervision should be to help people, machines, and gadgets to do a better job. Supervision of management is in need of overhaul, as well as supervision of production workers.
8. Drive out fear so that everyone may work effectively for the company.
9. Break down barriers between departments. People in research, design, sales, and production must work as a team, to foresee problems of production and in use that may be encountered with the product or service.
10. Eliminate slogans, exhortations, and targets for the work force asking for zero defects and new levels of productivity. Such exhortations only create adversarial relationships, as the bulk of the causes of low quality and low productivity belong to the system and thus lie beyond the power of the work force.
11a. Eliminate work standards (quotas) on the factory floor. Substitute leadership.
11b. Eliminate management by objective. Eliminate management by numbers, numerical goals. Substitute leadership.
12a. Remove barriers that rob the hourly worker of his right to pride of workmanship. The responsibility of supervisors must be changed from sheer numbers to quality.
12b. Remove barriers that rob people in management and in engineering of their right to pride of workmanship. This means, inter alia, abolishment of the annual or merit rating and of management by objective.
13. Institute a vigorous program of education and self-improvement.
14. Put everybody in the company to work to accomplish the transformation. The transformation is everybody's job.

Figure 12-1. Dr. W. Edwards Deming's management treatise can be the foundation on which productivity grows, in small plants as well as large corporations.

If everyone were to use this as the framework for their actions, there likely would be a reversal of the decline in jobs and the frustrations being exhibited today by all levels in industry.

It is my feeling that a sensitive monitoring system can enable a company to take advantage of changes instead of being thrown off course by them. In such a system, the worker is regarded as an important ally in the pursuit of quality improvements. The use of statistics to measure variations permits the early detection and correction of the process.

Taylor's methods may lead to increased productivity, but they have limits on what can be achieved. Deming's methods place no limits on what can be done. In Taylor's system, results seldom go beyond the imposed limits and often fall below them as soon as something causes an interruption or change. Deming's methods are much more flexible and depend on the ingenuity of people, combined with statistics for continuous improvement. The focus is on removing obstacles to quality.

I, along with many others in industry, am convinced that Deming's methods will better serve the manufacturer today. There is a need for speed in change due to new technology, competition, unpredictable world conditions, and higher demands for quality.

What we do not need is a quick fix. We have been looking too intently at single items to shorten time to market. We have come up with JIT, TQM, CAD, CAM, CAE, CIM, MRP, DFA, and other TLAs (three letter acronyms) that do give paybacks, but these fall short if mentalities aren't changed.

What the Deming plan tries to accomplish is cultivate a habit of improvements that never comes to an end. It calls for a profound change in how management operates and reacts from top to bottom — an entirely different management concept from the Taylor methods characteristic of most of this century here in the U.S.

Monitoring the process

Just as there is no intention in this book to provide a complete course on activity-based management or ISO 9000, the same is true for statistical process control. However, it is appropriate to point out the increasing necessity for managers to have a knowledge of SPC to effectively operate a paint shop now and in the future.

As a preface to outlining a quality control system, I wish to comment on a phenomenon emerging ever increasingly in finishing operations: the proliferation of control charts. It seems to be conventional belief that if walls are papered with control charts, it will either impress someone or cause the performance to improve. Usually, neither is the case.

It is possible today to collect so much information on a paint shop you could literally be buried in it. When you test how much of this is really understood, how it is being used, and how long it takes to react, it is obvious the potential for improvement is not yielding proper value. It is just charts for chart's sake. It would be so much better to have fewer charts and realistic time frames to interpret data and make quick adjustments to obtain real results.

This is another version of the "Ice Cream Principle," of Chapter 7. Some organizations are trying to eat the entire gallon instead of a sensible amount.

This is not to say statistics aren't needed. They are an integral part of a good quality control system. Figure 12-2 shows five items essential to a quality control system. First, certain requirements must be defined that identify the parameters of quality. Then there must be provision by staff for process monitoring, a set of procedures to follow, and the actual collection of data. Capability must be designed in to measure performance through various methods or systems and an inspection step must be included that contains both the how and the what we are doing. All these put into operation leads to improvement by problem solving.

If we break each of these five items into separate entities, a better understanding may be had. Figure 12-3, *Requirements*, will demand there be

---

Quality Control System

- Requirements
- Process monitoring, procedures, and data collection
- Measurement systems
- Inspection
- Problem solving

Figure 12-2. The five essentials for effective quality control.

a management commitment, and it should be at the highest level. In my generation, it was said, "You put your money where your mouth is." Today, people will say, "Walk like you talk." It means the same thing however you choose to say it. You live and do what you propose.

The value of training has been proven. This means we fund and dedicate sufficient time for training, have it conducted at all levels, and in selected areas such as the paint shop. It will require resources such as manpower and equipment and can only happen through conscious effort. A key person should be named to be responsible for the activity and performance. Some method of positive reinforcement and reward must also be established when improvements sought are made.

Figure 12-4 outlines the various parts of process monitoring.

The process flow diagram (Figure 12-5) shows the entire process, including all functions. A comprehensive chart such as the one shown leaves no doubt as to the elements of the process and the sequence in which they fall.

Requirements

- Management commitment
- Training
- Resources

Figure 12-3. *Quality begins at the highest levels and includes the entire organization.*

Process Monitoring

- Process flow diagram
- Failure mode and effects analysis
- Key process variables
- Standard operating procedures
- Data collection

Figure 12-4. *Total oversight of system processes provides a continuous motion picture of operations.*

The yield of non-defective pieces from a process is really the rollup of all defects from all process details in the process. If a 5% defect rate (95% good yield) were attained on each step over a simple 10-step process, only a rolled-throughput yield of about 60% would be achieved.

$$(0.95)10 = 59.8\%$$

Managing a Paint Shop

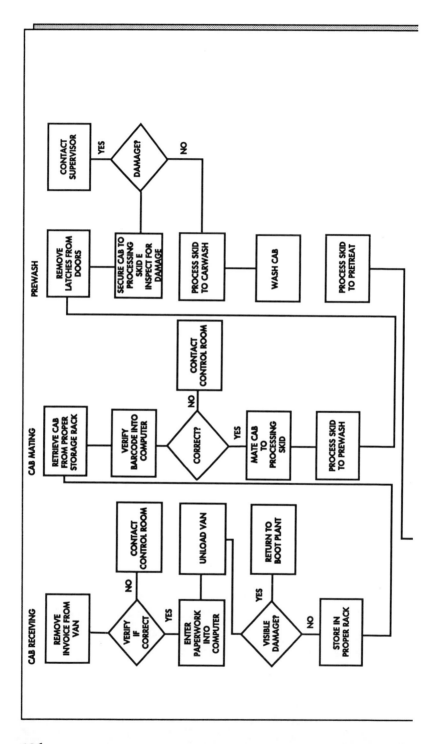

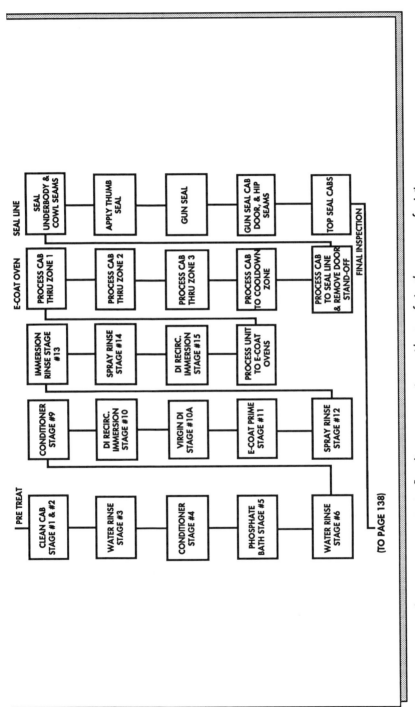

Figure 12-5. A comprehensive process monitoring flow diagram communicates without confusion the sequence of painting operations.

Managing a Paint Shop

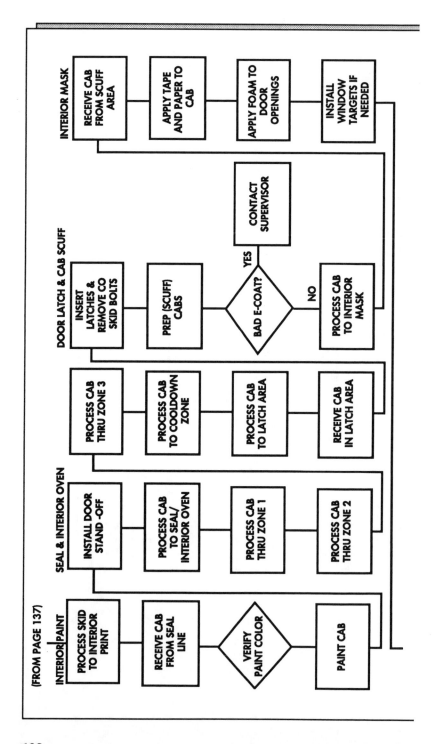

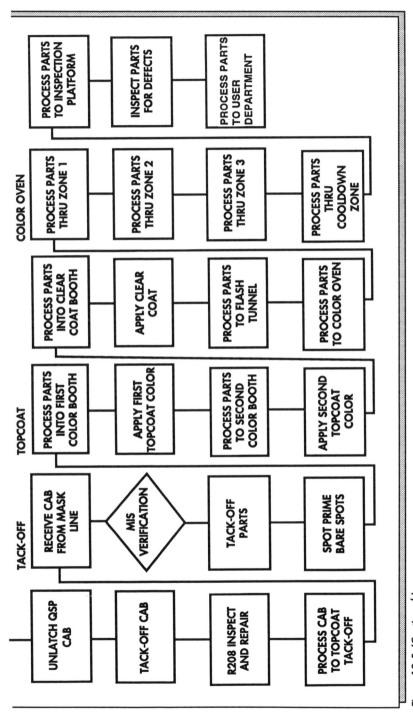

Figure 12-5. (Continued.)

If each of the process steps were to yield just 1% defect rate (99% good yield) and the process were a bit more complex (25 details), the total rolled-throughput yield would still be a mediocre 78%.

$$(0.99)25 = 78\%$$

From the flow chart shown, we can get an idea of how low a rate of performance could prevail if the same 1% defect rate were applied to each of the details listed. There are at least 80 process detailed steps in this illustration. This is not uncommon today in a high-volume, computer-controlled and automated paint operation. The fact that such paint lines operate at excellent yields is a testament to the capabilities of both man and machine. The point to be made is important to any manager of a paint shop: You must be intimately familiar with all the operations that affect your performance. That is the purpose of using correct process flow charts.

$$(0.99)80 = 44.3\%$$

Once again, how good is good and what is the quality standard? Let us raise the ante in this game of performance. You feel that a quality standard of 100% is just not going to be attained. After all, people are not perfect. So would it be better to recognize this and set our standard with a bit of that recognition in it? Let us set the quality standard at 99.9%. To be sure, that would be a pretty impressive record to show the boss. Or would it? Take a look at what 99.9% correct would mean in other scenarios:

- Two million documents will be lost by the IRS this year.
- 811,000 faulty rolls of 35mm film will be loaded this year.
- 22,000 checks will be deducted from the wrong bank accounts every 60 minutes.
- 1,314 phone calls will be misplaced by telecommunication services every minute.
- 12 babies per day will be given to the wrong parents.
- 268,500 defective tires will be shipped and installed this year.
- 14,208 defective personal computers will be shipped this year.
- 103,260 income tax returns will be processed incorrectly this year.
- 2,488,200 books will be shipped in the next 12 months with the wrong cover.

Where pride prevails, quality resides

Whether we want to admit it or not, there is only one course back to competitive excellence: 100% performance. If anything less is targeted, consider the ramifications of the alternatives previously listed.

The Malcolm Baldrige National Quality Award contains language that describes quality as a "race with no finish line." There is no period in a professional paint and finishing manager's career when it is okay to let up. Years and years ago, I created the chart at right on reliability (Figure 12-6).

Paint reliability is daily effort. Of course, the first letters spell PRIDE. Following the statements on the chart will bring pride to the product, people around you, the company, the customer, and yourself. The chart is over 30 years old, but the idea and its need is just as current as anything new being proposed today. Good ideas have that quality. They stand the test of time.

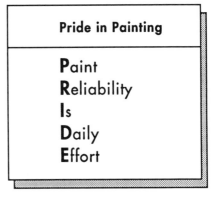

Figure 12-6. Key to productivity is pride in workmanship.

A colleague of mine had a chart on his office wall defining mathematically his approach to quality:

$$VALUE_p = \frac{Q_p \times S_p}{C \times T}$$

Where:
Q = Quality
C = Cost
p = Perceived
S = Service
T = Time

The point is, regardless of how quality is defined or which values are applied to it, the prevailing pathway to achieving it is persistence. You are not going to be competitive and a player in the game of business in the future if you do not lead the paint shop in the direction of quality every day.

A good quality program and efforts to resolve problems will require what is known as "Failure Modes and Effects Analysis," using a form such as that shown

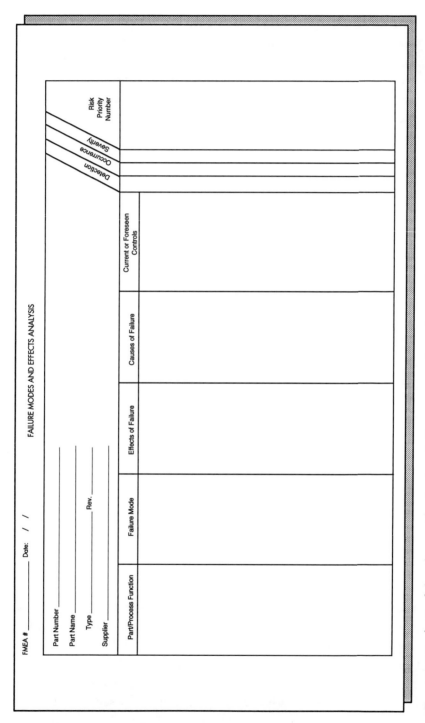

Figure 12-7. By charting failures from part function through effects, causes, and controls to remedy, managers are better informed to take corrective action.

in Figure 12-7. This method is used for tracking problems you may have with suppliers. The suppliers may be firms outside the company or other areas/divisions/plants of your own company. Being a part of your own company should never exclude the same consideration and rules for product quality as those imposed on outside suppliers.

Too often this is the case. We seem to live by a double standard in industry wherein outside suppliers are often dealt with more harshly than those within our own company. This was never fair or right, and it still is not, but it still happens.

There have been several occasions in which goals were set for improving incoming product quality, but these efforts did not include internal sources. The goals could not be met from the very beginning because certain basic information was either not provided or it was ignored. If every outside source was perfect, the combined effects of all internal sources would add up to a higher level than the goal. Each internal supplier had what was considered high quality. When the sums of all sources were compounded, the game was lost before it began. Yet, all individually could claim they met the standards.

It comes down to a matter of needs and wants. In the previous example, the paint shop got what it wanted — a product that had 5% or less defects from the suppliers — but what they did not obtain was what they needed to meet their performance goal of 95% first-time yield. Management did not comprehend the effect of cumulative defects and their potential reduction of first-time yield.

Figure 12-8 serves as an example of what might be required in your operation to identify and analyze a problem so corrective action may be taken.

To establish any form of measurement in the paint operation, it is necessary to have a set of standard operating procedures. These must be clearly defined, step by step, and it is very helpful to have visuals like drawings, pictures, or even videotapes.

Procedures are an aid to paint shop workers in recognizing the variables that can creep into any system. Unfortunately, few paint operations have such a set of SOPs. The influence of ISO 9000 and SPC techniques has

Measurement System

- Data must be reliable
- What is the accuracy?
- What is the repeatability?
- What is the reproducibility?

Figure 12-8. Measurement is key to problem analysis.

brought higher visibility to the need, but it is still a major shortcoming in most small paint shops.

Many of the variables affecting painting operations are listed in Chapter 11, Figure 11-1. There are some additional process variables which should be pointed out and are not included in a facilities review. There can be great differences in the *experience* of personnel. It could be argued this would come under training, since it was one of the listed variables. Experience is an asset that cannot be taught. It is acquired knowledge. You cannot train a performer in how to react to stage fright or someone missing a line or cue; experience, however, cultivates the composure to continue as if nothing unplanned had happened.

Suppliers can change and their performance can change, and this can create variables outside your influence, at least initially. The selection and monitoring of suppliers has come a long way in recent years, especially where large companies are involved. Thousands of suppliers to smaller operations simply do not meet the modern expectations of quality. This results in a great amount of wasted assets and stress to overcome this form of variation. Much is still to be done.

Engineering in the form of design changes, special features or equipment, and broken time promises for implementing changes are often disastrous for high performance. Common results are that the product cannot be cleaned or drained, the metal is too heavy for curing time/temperature or oven cycles, and correct hooks are not provided because there isn't any time.

Once there is a set of standard operating procedures and a good understanding of all the major variables expected potentially, there will be a need for good data collection methods. These will include check sheets, SPC charts, and use of computers to collect, store, and interface with other computers, such as mainframes.

Many smaller operations may not have all the computer resources of larger shops. While it is desirable to use such technology, many charts and data collection activities can be performed manually. A well prepared inspection chart will be adaptable for either method. It will serve as a valuable tool in tracking quality and reacting to problems so corrections can be made.

Figure 12-9 shows an "Inspection Work Card" capable of being used in either electronic or manual mode to record defects. It gives an outline of the product and where the defect occurred and identifies the defect by using the defect code provided. The various zones of the product are numbered to facilitate identification and location.

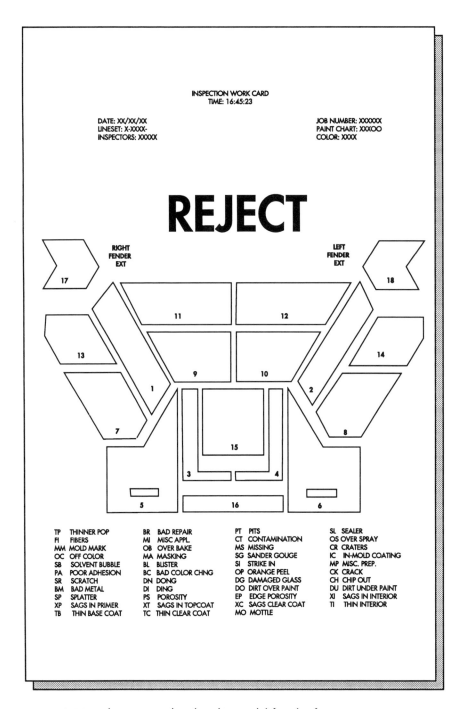

Figure 12-9. Visual inspection aids such as this speed defect identification.

The measurement system is of value only to the extent that the data derived from it is reliable and accurate. There must be a very high level of repeatability and reproducibility.

The inspection process needs to answer the question of *how* the inspection will occur. It must, of course, identify nonconformities in both the process and the product, but will this be done by a first-piece sampling, randomly, 100% inspection, or some combination of all in a differing needs scenario?

The other question needing an answer is the *what* side of the methods employed. Universally accepted standards should be communicated and fully understood. Physical examples of defects such as fisheyes or sand scratches displayed on panels or parts mounted on boards get the message across quite effectively. For other types of defects panels could display the acceptable standards based on the aid of equipment. These might include gloss or DOI examples.

In other cases, the use of drawings, photos, and videos could be used to confirm or verify if a process was being repeated properly and had not drifted over a period of time.

The end product of all this activity should be improved performance. If this is not the end result, then the only thing achieved was a rise in costs. Done right, quality will be improved and costs will drop. Problem solving is the payoff in a well managed quality effort. Everybody wins.

Figure 12-10 enumerates the steps that can effectively turn a problem into an improvement. These steps are largely self-explanatory, and each is critical to supporting the eventual change to come. It points the way to discipline and logical thinking that ultimately bring order. The nine-step sequence helps

Problem Solving

1. Identify problem/opportunity
2. Pick the team
3. Review/create process flow diagram
4. Brainstorm
5. Rank the variables
6. Set up experiments
7. Make changes
8. Monitor the process
9. Start over again at step 1

Figure 12-10. Order in problem solving relies on a logical sequence of steps leading to solutions.

prevent the "shooting from the hip" response that so vividly confirms the saying "that for every manufacturing problem, there is a quick and simple solution...that is wrong." Use of a fishbone chart such as that shown in Figure 12-11 will prevent such an approach and bring into focus all the variables that can impact on paint appearance. Your particular system may not have all the same variables, but the diagram gives a quick view of the types of entries included for consideration. They are either contributors to good or bad performance.

Is it good?

Managers must constantly ask, *How good is good?* Quality is a highly dynamic subject because it is constantly being redefined and improved. The former conventional wisdom purporting that you can't have quality and quantity is dead. In fact, there no longer is any justification to not have both.

Most quality systems in the past have been driven by tenets of Frederick Taylor. For many years, his theories made great gains and achieved dramatic results, but they no longer apply. To be competitive in today's manufacturing environment means practicing continuous improvement as suggested by Dr. Deming.

Where Taylorism has limitations on performance, the Deming method knows no bounds, based as it is on reducing complexity at all levels. The Deming approach deals with adjusting to change rather than denying its existence. The failure of American management to embrace such methods has led many to lose competitive edge, resulting in serious financial and social consequences. Others in the world have used such quality concepts to gain leadership and highly competitive positions.

ISO 9000 certification and SPC are spreading rapidly and becoming the basis for determining qualification for doing business with companies. The global impact of ISO 9000 will have far-reaching impact on most businesses in the coming years.

All organizations must make this new way of approaching problem solving and improvement a way of life. The old levels of 95% or even 99% first-time yields are inadequate. It can even be argued that 99.9% is not good enough. First-time yields of 100% are being achieved in some industries, and should be the benchmark for all of industry.

Toward that end, a fully dedicated management and a quality process is mandatory, and every person in the organization must be a player in the effort.

Managing a Paint Shop

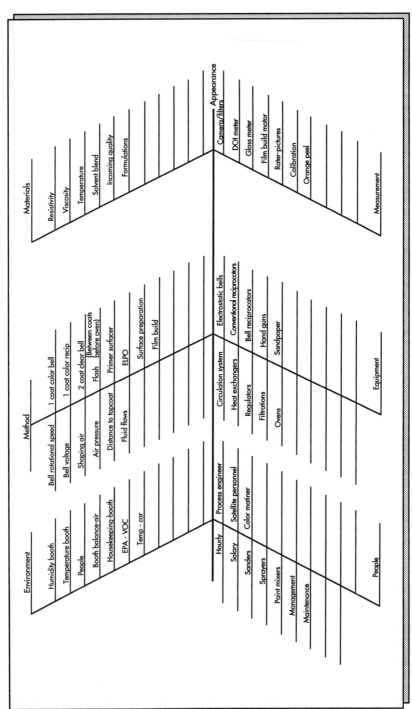

Figure 12-11. Charting the variables affecting paint appearance helps prevent hasty, ill-derived responses to problems.

Quality control must make provision for process monitoring, standard procedures, and data collection. In turn, measurement systems must produce data that is reliable and accurate. Repeatability and reproducibility of the process are major parts of measurement.

We are in competition with the world for supplying goods and services. There are few islands today. How well we perform in this competition will have significant impact on our quality of life as individuals and our future as a nation. If we continue to lose the industrial base we have created and enjoyed for the last two centuries, we may well lose our American way of life.

We can only spell *pride* if paint reliability is a daily effort. The Malcolm Baldrige Award describes quality as a "race with no finish line." That succinctly describes the subject, the need, and the challenge before us. Rework is waste, and waste is cost.

Chapter 13

Painters' Math

The best gallon of paint or pound of chemical is the one we don't have to use to get the job done properly.

Anyone involved with painting operations should have an understanding of the "painters' math" used to determine paint solids for area coverage and mileage, transfer efficiency, paint cost per piece, and VOC emissions.

Some years ago a good friend and colleague wrote a summary article for a trade journal addressing these issues. It is so well put together I find no reason to write another. With Jack Stauffer's permission, and that of *Products Finishing* magazine, I am proud to share in this book the following chapter.

Paint solids determination

When we were much, much younger, we worked as helpers to an old house painter who tried, sometimes in vain, to make us useful members of society. In hopes of making us into at least mediocre painters, he passed on to us a small, worn volume entitled "Math for Painters." It included tips on estimating coverage for various types of paint, as well as "recipes" for the amounts of white lead, oils, and turpentine to be used in concocting primers,

---

Reprinted with permission of the publisher, from the August and September issues of *Products Finishing*, trade magazine for the finishing industry, Copyright© 1984 Gardner Publications, Inc., 6600 Clough Pike, Cincinnati, OH 45244.

topcoats, and enamels. Alas, the book is lost now, along with other bits and fragments of the past, but we occasionally recall it as a tool as valuable as some of the costly brushes we used then.

Modern production painting has its own math. Fortunately, we can leave the compounding formulas to the suppliers of the paint. But we ought to be able to handle the math relating to mileage, losses, transfer efficiency, cost per piece, and solvent emissions, if we are to get our money's worth out of the sizeable investment our company has made in equipment, facilities, manpower, and materials. As importantly, we need to be conversant with formulas for determining volatile organic compounds (VOCs), in order to comply with EPA regulations.

Let us define the important factors relating to paint mathematics and shape them into formulas useful in obtaining answers that will tell us about the health of our finishing system.

Percent solids. When we buy a gallon of paint, we receive two things: solids and volatiles. The confusing thing is that the solids are not solid, and at best, we can only smell the volatiles.

Solids are considered to be the portion of the paint that will dry or cure to a solid film. They include the resins, fillers, and pigments remaining when the paint is dry or cured. The balance of the paint consists of solvents that evaporate during spraying, air drying, or curing. Since 100% of a gallon of paint must be something, we usually say it is divided between solids and volatiles. Thus a paint that is 40% solids has 60% volatiles.

But it's not quite that simple. We can determine the percent solids of a gallon of paint most simply by heating it until all the volatiles are gone and comparing the weight of the remaining solids with the original weight. This will give us the percent of the original weight that is solids, or percent solids by weight. This is a handy figure to have when you are making paint, but won't help us much with our "painters' math," as we do everything in gallons, not in pounds.

A better expression of solids content is percent solids by volume, or "volume solids." The difference here is that we can find out how many cups, quarts, or ounces in the gallon are solids or volatiles. These can easily be converted into cubic inches and square feet covered.

The derivation of volume solids starts with weight solids and modifies these figures by the specific weights of the various components to arrive at volume. Percent volume solids allows us to determine how many of the 231 cubic inches in a gallon are useful film formers and how many of them will evaporate.

Thus a gallon of paint that is 40% solids by volume has 231 x 0.40 = 92.4 cubic inches of useful paint solids.

Paint volume solids range from as low as 10% for some wood sealers to as high as 100% for radiation-curable coatings. The definitions of "low solids" and "high solids" are still indistinct but the dividing line is usually accepted as being between 60% and 70%.

How do you find the volume solids of the paint you are using? Ask your supplier. He has the numbers or will get them for you.

One further mystery must be cleared up before we can claim mastery of "solids." Most paint is thinned or reduced after we receive it. This allows us to vary the solvent content and type to cope with weather, surface requirements, and varying spray equipment, as well as allowing us to buy our solvent locally and pay freight on fewer drums of paint. How can we know the volume solids of the material we are actually spraying? Refer to Table 13-1.

Mileage. When you bought your last car, the sticker on the window showed the estimated mileage as so many "miles per gallon." With paint, the mileage is usually expressed as pieces per gallon. When we use the terms square foot per gallon or square inches per gallon, we will refer to them as "coverage," to avoid confusion. The window sticker on the car qualifies the mileage figure as either "city driving," "country driving," or "average conditions," and then confuses the whole issue with a reference to driving habits.

At first, paint mileage seems as confusing as gas mileage, but with paint, we can identify and quantify the variables.

Table 13-1. Determining Volume Solids, Reduced Material

$$S_c = \frac{V_u \times S_u}{V_u + T}$$

Where: V_u = Volume of unreduced (uncut) paint
S_u = Percent solids, by volume, of uncut paint
T = Volume of thinner or solvent added
S_c = Percent solids, by volume, of reduced (cut) paint

Note: When percent solids is used in any of our formulas, we use it as a decimal (60% = 0.60).

From our previous discussion of volume solids, we would logically deduce that the more solids contained in a gallon of paint, the more pieces we could cover with that gallon. Thus the first mileage variable is volume solids.

The second mileage variable is film thickness. A dry cubic inch will cover one square inch (6.5 cm²), if the film is to be one inch (25.44 mm) thick. The same cubic inch will cover 1000 square inches (6500 cm^2), if the film is to be one thousandth of an inch (0.025 mm) or 1 mil thick. Remember, when we talk of solids, we refer to dry film thickness.

The third factor affecting paint mileage (and coverage, as well), is transfer efficiency. No application process is 100% efficient. Some of the paint is always lost as overspray, drips, and excessive film thickness. We will deal with determination of transfer efficiency later. For the purpose of mileage determination, however, consider transfer efficiency as a percentage, used as a decimal in our formulas. Using the above three variables, the formulas for coverage and mileage are as shown in Table 13-2.

These formulas are useful when comparing the performances of two paints or two types of application equipment. However, the actual mileage (M_u) on your production line can be measured. When used with later formulas, this can be the key to determining transfer efficiency and cost per piece.

To determine mileage on your line, it's best to make a test run. The run of the part in question should be as long as possible, to minimize errors in measurement. Paint usage during the run can be quite accurately measured by the following method:

Table 13-2. Determining Coverage and Mileage

$$C = \frac{1604 \times S \times E}{t}$$

$$M_u = \frac{1604 \times S \times E}{t \times A}$$

Where: C = Coverage in square feet per gallon
S = Volume solids, unreduced paint
E = Transfer efficiency
t = Dry film thickness, mils
M_u = Mileage, pieces per uncut gallon
A = Part area, square feet

At the start of the test run, measure from the top of the drum, bucket, or pressure pot down to the surface of the paint. At the end of the run, do the same. The difference in these two measurements, in inches, can be substituted for "h" in the formula shown in Table 13-3.

Let's assume, for example, that a paint container is 13 inches (330 mm) in diameter and that the first and second readings were five inches (127 mm) and 18-1/16 inches (472 mm). Thus "h" would be 13.06 inches (332 mm). Substituting in the formula in Table 13-3, we would have:

$$U = 0.0034 \times 13^2 \times 13.06$$
$$= 0.0034 \times 169 \times 13.03$$
$$= 7.5 \text{ gallons (28.5 L)}$$

Let's also assume that during the run, 645 pieces were painted. The mileage "M_c" in pieces per cut gallon (remember that the paint measured above has been thinned, or cut, and is not the same in solids as the paint discussed in the earlier formulas) can be obtained as shown in Table 13-4. Substituting for "P" and "U," we find:

$$M_c = 645 \div 7.5 = 86 \text{ pieces per cut gallon}$$

Table 13-3. Determining Amount of Paint Used From a Cylindrical Container

$$U = 0.0034 \, D^2 h$$

Where: U = Paint consumed, gallons
D = Inside diameter of paint container, inches
h = Distance between first and second readings, inches

Table 13-4. Determining Mileage in Pieces Per Cut Gallon, Based on Test Run

$$M_c = \frac{P}{U}$$

Where: P = Production, pieces painted during test
U = Paint consumed, gallons

It would be convenient to "plug" the value of M_c into formula 3, above, and solve for transfer efficiency; but formula 3 deals with uncut gallons (M_u). This brings us to the next formula, shown in Table 13-5.

For example, let's assume our paint is thinned before use by adding five gallons (19 L) of solvent to 40 gallons (152 L) of paint. Then:

$$M_u = \frac{40 + 5}{40} \times 86 = 96.75, \text{ or } 97 \text{ pieces per uncut gallon}$$

The fact that we always paint more pieces per uncut gallon than per cut gallon takes a little "getting used to" for some people. Just remember that a cut gallon is thinner, or has less solids, than an uncut gallon. So, it should cover fewer pieces.

Try it! Review these formulas and apply them to your situation. Make one or more "mileage" runs.

Table 13-5. Determining Mileage: Pieces Per Gallon, Unreduced

$$M_u = \frac{V_u + T}{V_u} \times M_c$$

Where: M_u = Mileage, pieces per unreduced gallon
V_u = Volume of uncut paint, mixed with
T = Volume of solvent used in mixing
M_c = Mileage, pieces per cut gallon

Table 13-6. Determining Transfer Efficiency

$$E = \frac{M_u \times t \times A}{1604 \times S}$$

Where: E = Transfer efficiency, as a decimal
t = Dry film thickness, mils
A = Part area, square feet
S = Volume solids, unreduced paint, as a decimal

Transfer efficiency

Since formula 6 (Table 13-5) has provided us with mileage in pieces per uncut gallon, we can use the result in an inversion of formula 3 to solve for transfer efficiency, as shown in Table 13-6.

In the formula shown in Table 13-6 we can handle dry film thickness in two different ways. First, we could measure coating thickness on say 10% of the parts painted, taking 10 readings on each part and averaging all the readings to arrive at a value of "t." Use of this method assumes the process is in control and that dry film thickness is within limits. Let's assume specs for the part require 1.2 mils plus or minus 0.2 mil, or 1.0-1.4 mils. Assume the average of all of our readings was 1.7 mils. If we use the 1.7 mils figure in the formula, we will calculate the efficiency of the painting device putting paint on the part. But is this really total process transfer efficiency?

Some would say that since any film greater than 1.4 mils represents waste, we should use 1.4 in the formula. If we did that, however, and the actual deposited film were 1.0 mil, we would be stating the efficiency as higher than it actually is.

Others would say that we should show the higher efficiency, since we are able to hold to the low side of the tolerance while maintaining quality.

As usual, both sides are right. If we are comparing different application methods, we would use actual applied film thickness. To measure overall process efficiency, we would use the nominal film thickness. Let us assume the latter is our need and use 1.2 mils.

Part area is the measure of the area of the part to be painted, in square feet. If square inches is a more convenient measure, substitute the number "230,976" for the number "1604" in formula 7 (Table 13-6). The area figure should represent only that portion of the piece requiring paint.

If electrostatic equipment is in use, "wraparound" is beneficial only if both sides of the piece must be painted. Otherwise, the area of one side only should be used.

Let us assume an area of 2.0 square feet for our example. The volume solids of our unreduced paint has been determined from our supplier to be 48%. Substitution in the formula may be made as follows:

$$E = \frac{97 \times 1.2 \times 2.0}{1604 \times 0.48} = 0.30 = 30\%$$

Be prepared for rather low efficiency figures and be wary of high ones. Nonelectrostatic air spray rarely exceeds 30% transfer efficiency (TE), and on small parts, often shows less than 10%. Nonelectrostatic airless spray shows slightly higher TEs. Electrostatic versions of these two processes should about double their TEs. Electrostatic bells and disks should show TEs in the 80% to 90% range.

Measurement of transfer efficiency, while not showing actual painting cost, is very important. It shows our losses and our potential for improvement.

Let's assume our total annual expenditure for paint and solvent is $40,000, and we have measured our efficiency as above, at 30%. This means we are putting $12,000 (40,000 x 0.30) worth of paint on our parts, and throwing away $28,000 per year. Thus our target for cost reduction is $28,000 per year. A doubling of transfer efficiency to 60% would still put $12,000 worth of paint on the parts per year, but at a total cost of only $20,000 (12,000 ÷ 0.60), for an annual saving of $20,000. Just as each of us should have an annual physical exam, our paint line should get a TE check each year.

Paint cost

Thus far, we have solved all the mathematical mysteries of our paint line except one. To find the actual cost of paint and solvent used, per piece, we can use the formula shown in Table 13-7. For example, let's assume a paint cost of $11.00 per gallon and a solvent cost of $2.00 per gallon. (If more than one solvent is used, a cost-weighted average must be used.) We will use the reduction ratio (40 gallons paint, and 5 gallons solvent) and M_c (86 parts per cut gallon) from our earlier example. Then:

$$X = \frac{(40 \times 11) + (5 \times 2)}{86(40 + 5)} = \$0.1395 = \$0.14 / \text{piece}$$

Table 13-7. Cost of Paint and Solvent Used Per Part Finished ($ Per Piece)

$$X = \frac{V_u \times C_u + T \times C_T}{M_c (V_u + T)}$$

Where: X = Cost, in $ per piece
V_u = Volume of uncut paint used when mixing
C_u = Cost of uncut paint, $/gallon
T = Volume of solvent used in mixing, gallons
C_T = Cost of solvent, $/gallon
M_c = Mileage, pieces per cut gallon

Emissions

In many locations, it is important to know (and to report) emission of VOCs from the paint system. Solvent emission in gallons per piece, or per thousand pieces, is seldom a useful figure, however. Few paints contain only one solvent when reduced for use, and few plants process only one part for a significant length of time. Governmental agencies will normally be interested in the *weight* of VOCs emitted per unit of time (day, month, or year).

Two different approaches to VOC calculations are used. One is historical, in that it addresses actual emissions, measured over a past period, for an existing source. The other deals with estimates for a future time period, either for an existing source, a modification to an existing source, or a planned new source. The methods used for the two calculations are vastly different and will be discussed separately.

Calculation of actual emissions for a past period is based upon paint and solvent usage records for the period. VOC emissions from a paint system come from three sources: evaporation of solvents in the paint as received from the supplier; evaporation of solvents added to the paint by the user; and evaporation of solvents used in the system for housekeeping or cleanup.

Your materials or purchasing group is the best source of figures for gross consumption of paint during a period of time. Total purchases for the period and beginning and ending inventories will yield this figure. As mentioned earlier, your supplier can provide you with the percent solids for your coating materials. In this case, since we are interested in emissions in *pounds* of VOCs, we will use the *weight* solids. To find the weight of solvent in a volume of paint, we use the formula shown in Table 13-8.

Table 13-8. Determining Weight of Solvent in Unreduced Paint

$$L = V_u \times W \times (1 - S_w)$$

Where: L = Weight of solvent in uncut paint, lb
V_u = Paint volume, uncut gallons
W = Weight of paint, lb/gal (from supplier)
S_w = Percent solids, by weight (from supplier)

Emission of solvents used for reduction and cleanup can be determined by a similar approach. Again, your purchasing or materials group can provide total usage for the period. This total, less dirty solvents shipped out as scrap, is emitted from the system (by evaporation) from the spray booths, flash off area, and bake oven or drying area. It must be counted as discharge of VOCs. Total weight is calculated as total volume of each solvent times its weight per gallon.

Calculation of emissions for a future period is in fact a forecast. As such, it must be based upon production forecasts for the period. Your plant's data-processing group should be able to generate a schedule forecast of parts to be painted for the period, with quantities of each.

Table 13-9. Determining Amount of Solvents in Paint; Amount Used for Reduction

$$L = \frac{V_u \times w \times (1 - S)}{M_c (V_u + T)}$$

$$R = \frac{T \times w}{M_c (V_u + T)}$$

Where: L = solvent in paint, as received, lb/piece painted
V_u = Volume of paint to be reduced, gallons
w = Solvent weight per gallon, lb
S = Paint volume solids, percent
M_c = Mileage, parts per cut gallon
T = Solvent used in reduction, gallons
R = Solvent used in reduction, lb/piece painted

Table 13-10. Finding a Constant Applicable to All Parts Produced: Solvents in Purchased Paint Solvents Used for Reduction

Let: $$\frac{V_u \times w(1-S)}{V_u + T} = f_l$$

and: $$\frac{T \times w}{V_u + T} = f_r$$

then: $$L = \frac{f_l}{M_c}$$

and: $$R = \frac{f_r}{M_c}$$

As with historical calculation, we must deal with solvents received in the paint, solvents used for reduction, and cleanup solvents. The first two factors can be calculated in pounds per piece painted, using the formulas shown in Table 13-9, based upon mileage (M_c) in pieces per cut gallon. The third factor (cleaning solvent) must be estimated, determined from usage records, or both.

The values for "L" and "R" can be multiplied by the production for each part in the forecast to determine the total emission in pounds for the time period. Although it would seem a formidable task to make the above calculations for the many parts processed through a typical finishing system, many of the factors can be resolved into a constant common to all the parts produced, as shown in Table 13-10.

The above emissions calculations assume that all the solvents in question are hydrocarbons subject to regulation. Water-borne paints can be used in the calculations, but the water must be deducted, as it is exempt. Also, the chlorinated solvents 1,1,1-trichloroethane and methylene chloride may be exempt in your locality and can thus be excluded.

All formulas and factors are summarized in Table 13-11 as a convenient reference.

Table 13-11. Summary of Formulas Used for "Painters' Math"

1. Volume solids, reduced material:

$$S_c = \frac{V_u \times S}{V_u + T_c}$$

2. Coverage, square feet per gallon:

$$C = \frac{1604 \times S \times E}{t}$$

For square m, substitute 144 for 1604.

3. Mileage, pieces per gallon:

$$M = \frac{1604 \times S \times E}{t \times A}$$

4. Paint consumed from a cylindrical container:

$$U = 0.0034 \, D^2 h$$

5. Mileage, pieces per gallon, based on test run:

$$M_c = \frac{P}{U}$$

6. Mileage, pieces per gallon, unreduced paint:

$$M_u = \frac{V_u + T}{V_u} \times M_c$$

7. Transfer efficiency:

$$E = \frac{M_u \times t \times A}{1604 \times S}$$

8. Paint material cost, $ per piece:

$$X = \frac{(V_u \times C_u) + (T \times C_T)}{M_c (V_u + T)}$$

9. Solvent weight in unreduced paint:

$$L = V_u \times W \times (1 - S_W)$$

10. Solvent weight in unreduced paint in pounds per piece:

$$L = \frac{V_u \times w \times (1 - S)}{M_c \times (V_u + T)}$$

11. Solvent weight in reduction, pounds per piece:

$$R = \frac{T \times w}{M_c \times (V_u + T)}$$

Where:
- A = Part area, square feet
- C_T = Solvent cost, $/gallon
- C_u = Cost of unreduced paint, $/gallon
- D = Inside diameter of paint container, inches
- h = Height difference, paint used from container, inches
- L = Solvent in paint, as received, lb/piece painted
- P = Production, pieces painted during test
- R = Solvent used in reduction, lb/piece painted
- S = Volume solids, percent
- S_W = Weight solids, percent
- t = Dry film thickness, mils
- T = Volume of solvent used in paint reduction, gallons
- V_u = Volume of paint used in paint reduction, gallons
- w = Weight of solvent, lb/gallon
- W = Weight of paint, lb/gallon
- U = Paint consumed, gallon
- M_u = Mileage, pieces per uncut gallon

CHAPTER 14

Problem Solving and Decision Making

Paint shops are for finishing products, not rebuilding them. If the rest of the plant is permitted to flush its shoddy work down to the paint area, then we become a waste treatment facility, not a finishing operation.

Problems are in the eye of the beholder

At one time or another we have all probably heard that a problem is not a problem but an opportunity. In my case this was how my early training was directed and how my mental processes were molded. Problem solving has always been an upper for me, a real challenge.

This was further reinforced on an occasion some years back. A group of paint people were attending a training session in Muncie, Indiana. The one-day seminar was long, and it continued well into the evening hours after dinner.

By the time most had finished eating, the long day and sated appetites began to take their toll as a number of the participants began to fade a bit. The instructor, a one-of-a-kind sort of fellow by the name of Barney Sincock, was about as round as he was tall and had a uniquely unforgettable voice. He knew he had to do something to get everyone back to the business at hand.

Barney stood at the front, reared back, pulled open his coat, and bellowed, "I've been in the paint business nearly 35 years and have never had a complaint!" Now this brought a number of people back out of their stupors and up on the edges of their seats. Responses such as "Where do you get off

163

saying that? Don't you remember the time down at XYZ Company we had the black jell in the lines?" came from all corners of the room as the revitalized audience recounted incidents from their careers.

Following several of these responses, Barney replied, "Yes, I know all those things happened. I said I had never had a complaint. Those things were not complaints. They were compliments! Anytime someone has a problem and calls you, that is the greatest compliment you can receive. It means they believe you can help them. They believe you know something more about the situation than they do. When people come to you in time of trouble, it is not a complaint. It is a compliment! And I've had a heck of a lot of compliments in my time!"

This experience alone has caused me to view problem solving in a very positive way. I was to tell this story all over the world because of the message it conveyed. There were some places where phones would ring off the hook, and no one would answer for fear of being eaten alive by the person on the other end. My comment to these folks would be, "Hey, there is a compliment probably coming in for you." After that, the phone got answered quickly.

An ounce of prevention...

On the other side of problem solving is another familiar adage: The best way to solve a problem is to prevent it. It is an old-fashioned philosophy, but it still is the best way to manage and operate. Out of this adage came a series of paths, all leading to the goal of prevention. As one might expect, the better everyone got at prevention, the fewer the "compliments" that came over the phone.

The lesson from Muncie is a valuable one, and it has contributed mightily at times to help take the fear out of problem solving. Experience taught me, however, that something was not all it should be even with that approach. It was a version of reaction-style management. It had to have a problem to begin with. Reactive problem solving always carries a higher level of pressure and risk to make the correction. Proactive-preventive-management is much more efficient.

Prevention became more and more a part of my efforts during the past three and a half decades of my career. It is driven by a deep desire to have detailed knowledge of the products, markets, people, processes, and materials, and all the alternatives potentially available to prevent problems.

Throughout this book, I have related much of what I have learned in my travels in the paint business. Now I'd like to explore a few other truths directed at problem solving and obtaining first-time yields.

Truths in painting

One of these is, *We finish in paint areas.* It sounds pretty straightforward. What it really means is we do not rebuild a product in the paint shop. The paint sprayer has no magic that will somehow correct all the manufacturing flaws that have occurred ahead of the paint shop. As in the computer world, garbage in, garbage out.

This is why good paint and finishing begins way back in the manufacturing operation; actually even before that, in the design stage. If the rest of the plant is permitted to flush its shoddy work down to the paint area, then we become a waste treatment plant, not a finishing operation. The product that comes from such an operation is not what is intended to move through the paint shop and out into the hands of the customer.

I have always insisted that such products be turned around and sent back to wherever it was not properly produced. Unless this happens, no one will get the message that you are the next customer in line, and you are not taking delivery. Of course it takes a bit of courage to do this where it has never been done before. It is a test of top management and their true support of quality. I have seen managers really incur the wrath of up-line workers and management for such an action, while at other times they have been backed up for their action. Some very remarkable improvements have been realized when "paint" problems got corrected where they should have been.

Yesterday's methods yield yesterday's results. This comes about through *inflexible thinking.* Once again, this is a common sense observation. If we continue to do something the same way, we are going to get the same results. Why is this so difficult to understand?

Possibly, people are either reluctant to make changes for a number of reasons, or their knowledge and quality of information does not warrant such actions without great risk. Or, perhaps, the system that permits change is burdensome and unyielding. The bottom line is, when you view all the potential variables that can occur in today's world, the thing needed least is inflexible thinking. Response to change is playing an ever-increasing role in successful management.

In one of my most recent projects, a company had been a part of the military supplier industry. They produced a very fine product, considered to be the benchmark product of its kind in the world. With the reduction of Pentagon funds and downsizing, it was apparent they could not function in the future as a viable entity based upon military business alone. Something had to change, and quickly.

This company then purchased a well known older company that was in bankruptcy and had ceased operations. The military supplier company had a great deal of talent and a spirit to survive so often lacking today. Within 180 days of acquiring the bankrupt company, the parent organization had moved the major manufacturing facilities and equipment, converted their existing operation, retrained its personnel, set up a marketing group, and sold and built the first product. The customer accepted delivery with the statement, "This is the best unit ever received from this brand name, and it is only your first one."

This is problem solving at a high level, with corporate existence at stake. There was a problem of diminishing military business and a possibility of going out of business. There was a problem of a once proud name now having to be disposed of through bankruptcy. Through astute management, one company found new markets, another was rescued from dissolution, and the consolidated new corporation emerged strong, healthy, and sound. Problem solving does not always have to relate to the more narrow areas of improving a process or repairing a tool. Every effort toward continuous improvement is a form of problem solving or changing of yesterday's methods.

For every manufacturing problem there is a solution that's quick, simple, and wrong. How true! It is so tempting to make a quick fix for the moment rather than search out the real culprit and make the proper correction. Typical are these real shop-floor examples.

- Time after time, paint organizations experience problems like fisheyes caused by silicone contamination. Instead of looking for the source of the silicone, paint managers invariably opt to treat the paint with fisheye eliminator. Once these products are introduced, it is a major effort to stop using them. The product no longer exhibits the problem of fisheyes, but there may now be problem of masking tape not wanting to adhere. It may adhere initially, but blow off easily in the ovens and cause defects in the paint from sticking. Managers treated symptoms, not causes.

- In a slightly different example, I have seen managers try to save a few cents on such things as "O" rings by buying them from a generic supply catalog instead of the equipment supplier. Without warning, leaks popped up everywhere from guns and quick connections because the new "O" rings could not withstand the solvents used in the paints.

- The storeroom runs out of filters for the paint booth. Someone places an order for a local purchase, and replacement filters are received. But filters are not universal. Furnace filters were delivered instead of paint system filters and fibers speckle the paint. No one can understand the number of rejects. And, instead of tracing down the cause, management is chasing down the culprit because new filters were just put in.

These are just a few examples of quick fixes gone bad. Such events occur by the hundreds every day. They are seldom costed or even reported for fear of punishment or embarrassment. There are many ten-thousand-dollar mistakes that go unreported. Everybody feels now that this has happened, they won't let it happen again. Unfortunately, people move around on jobs, and new people may make the exact same mistake sometime in the future.

A grave error can be made by *confusing seniority with capability*. With technology changing many of our processes and materials, it is necessary to recognize that our more senior personnel may no longer have superior problem solving capability. Often, a younger person with less service has specific experience in new technology and may have better problem-solving and reaction skills.

It is always necessary, therefore, to ensure that new technology or any change is accompanied by retraining of key personnel and their backups. There have been cases of gun repairmen who had been working with electrostatic guns for years perform poorly when new guns were introduced. The older guns had separate power units outside the paint booths. There was only one air adjustment, and that controlled both the power unit and the atomization. The new gun had a built-in turbine instead of the more conventional power unit and there was a separate air adjustment for the turbine. This adjustment was not made initially, and the result was that turbines wore out rapidly, losing the electrostatic efficiency and causing several problems. Not only did this cause undue expense of replacing the turbines, but a great deal more paint was consumed. Rework from poor hiding and edge covering also resulted. Yet, only the most senior people were available to perform what appeared to the less trained people to be the same work. Management did not have deep knowledge of the equipment.

Contingency management

Despite the best efforts to prevent problems, there almost always comes a time when you need correction, and fast. A number of aids are available to help in this effort. Good records and daily logs of events top the list. These are forms of information. Nothing can compare to the capability of a well informed, well trained and experienced person with the mental attitude to handle the pressure. The touch of the master's hand is often the difference between success and failure in problem solving, but such aptitude and talent is not readily at hand.

Those who do possess it have the ability to take charge in a crisis, direct personnel, bring calm, and restore production with a minimum of time and money. Their abilities come from a mixture of intuition, experience, and proven theories. They are an encyclopedia of information who can turn their knowledge into performance. They are invaluable to an organization, especially when a line or system that is down can add up to several thousand dollars a minute in losses.

In the rare instances when a major line or plant stoppage occurs, the tortoise-and-hare lesson doesn't really apply. Continuous improvement may have prevented the stoppage but it can't be used to fix it. Situations do not permit a slow and steady response to such conditions. Because such situations are unexpected, there has to be a much more rapid reaction, in which specially trained and experienced people must be called upon. Like insurance, these people are your operational parachute.

The problem is that these people are getting harder and harder to find. Some of our most brilliant young minds no longer seem to pursue this type of career. They prefer instead jobs where they can aspire to steady hours, fine offices, and few sudden demands or unsettling events in their lives. Somehow, becoming an expert troubleshooter in the plant does not carry the stature and prestige they seek. To them, the idea of working all kinds of hours in a plant at night is not why they spent four to six years in college.

There is also the issue of protecting human resources. There was a time when no company laid off the good troubleshooter. That person was just too valuable. This is not always the case now. With new emphasis on electronic monitoring, preventive maintenance, and predictive maintenance, some managers feel secure in the capability of technology.

They attempt to solve such on-line problems with lower levels of technical personnel. Suppliers are expected to support such needs, and outside services are sometimes contracted. Some just take the risk that no such need will be required and use a consultant when the situation occurs.

These approaches have produced all levels of results, ranging from excellent to very poor, depending upon a number of factors. On large accounts permanent personnel will likely be stationed in-house or someone is retained and on call. But when a plant is bleeding to death, it can't wait for someone to show up. A small account may never see help arrive.

Consultants are seldom called until there is a major problem. Most companies have exhausted both their in-house capability and the supplier capability before bringing in an independent consultant. Even then, it is possible to find not all consultants are equally skilled in going directly to the floor of a plant and resolving issues promptly.

This is not to say all these people do not have talent. All plants have unique facility and equipment configurations and methodologies for running them. These experts have to be aware of what they are looking for and trying to fix. Doctors come in many varieties, and it is important to know which ones are ministers, professors, brain surgeons, or engineers. Does the situation call for a prayer, a brain surgeon, or a doctor of resinology? When your brain is involved, you want the right doctor.

A book has been prepared to help smaller plants prevent, quickly recognize, and then solve paint defects. Since most paint operations in the U.S. still use liquid coating, the book, *Liquid Paint Finishing Defects*, focuses on liquid defects to better serve more managers. Available from the Society of Manufacturing Engineers, it is written for the shop-floor worker, not academicians, and is a valuable tool for the paint shop.

Another need a manager will encounter in preventing or solving problems is that of devising a systematic approach to planning and major decision making. Such an approach is a tool good managers use on a regular basis.

Brainstorming decisions

Few managers really recognize that they simply do not have good decision-making skills, and if they do, it is a very hard thing to admit. They get along using only the reasoning experience has given them. It takes a humble manager to admit that his or her decision-making abilities range from not very good to poor. Changing reasoning processes is difficult without some specialized training.

The Kepner-Tregoe training system centers on bringing several experienced managers together on a common problem. It is difficult to know how a single manager working alone is thinking, but in a problem-solving conference, the joint deliberations can be easily observed, recorded, and even quantified. The word "problem" can mean many different things to each of the managers when discussing the same issue.

The raw material of management is information. It is the grist of what we all work with, whether generated internally or externally. We have to know what is wrong before we can begin to correct a problem.

Management learns from its experiences. The mistake management makes is that they believe each other knows equally well how to use information. This is certainly not the case, because some people are much more skilled at using information than others. Drs. Kepner and Tregoe found seven basic concepts in problem analysis and seven in decision making by observing tactics of practicing managers . With these fourteen concepts, they concluded, a two-part cycle is created. One half is for problem analysis, and the other half decision making.

Figures 14-1 and 14-2 illustrate the two halves step by step. For a greater familiarization of at least one means of systematic problem solving and decision making, take one of the Kepner-Tregoe formal training sessions.

This method has been used in a very wide and productive series of decisions to solve various problems in paint areas. Everything from entire paint shops to various sources of individual supplies have been used as subjects for this form of decision making.

It also takes a certain kind of manager to submit to such a method of shared decision making. As you are part of a larger group, your vote is only one of several equal votes when final acceptance is sought. Many old-school managers do not buy into this, but instead feel this is a dilution of their authority and do not wish to have to participate in an open manner.

I once took part in such a group decision-making effort on one of the most important programs of my career. More than 30 persons were involved in the joint study from every discipline in my company. I was the Corporate Manager of Paint, but had only 1/33 of the vote — only approximately 3% of the vote, but every one of us was committed to doing what was best for the company. That's what made it correct. I had 3% of the correct joint decision, not 100% of a possible flawed decision.

There was a time when I viewed the use of teams or joint decision making as very suspect. Too many times such efforts came from weak management who did not have the capability to make critical decisions, with the idea being to share the blame in the event something did not perform well. If it worked, they would take credit for being so democratic in their management style. If it did not, they thought there was a safety blanket to hide under. Either way, the accepted wisdom was that there were benefits to be gained by operating in this manner.

It was only after being exposed to formal training in reorienting my reasoning approaches that this feeling began to change. There were managers who still did team decision making, but not the way it should be done. The more I participated in such sessions, the more it became evident that such a

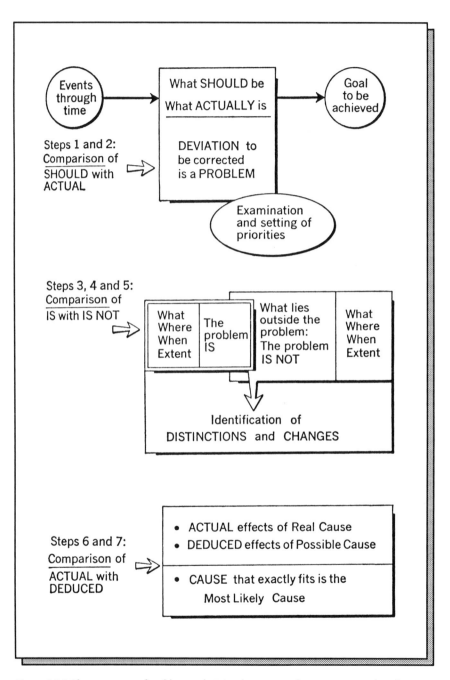

Figure 14-1. *The seven steps of problem analysis involve a series of comparisons made with various parts of the information available. These comparisons are made in order to arrive at a cause-and-effect conclusion that is the most tenable one possible.*

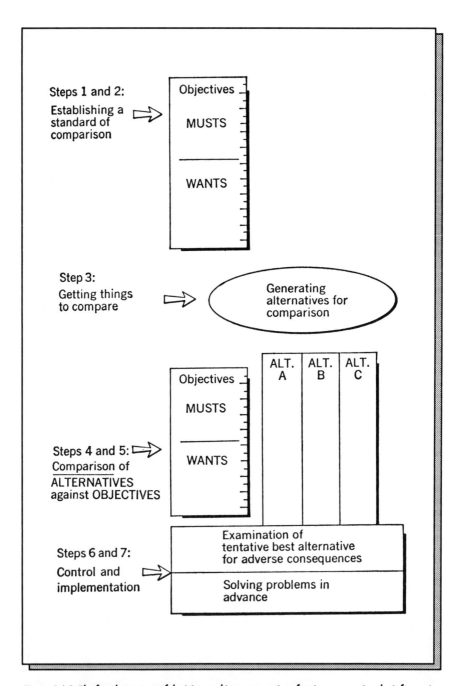

Figure 14-2. The first three steps of decision making are a series of actions preparing the information for the making of one complete comparison. This will determine the best action to take on a problem. The last two steps compare the expected results of this action against its possible adverse effects.

method caused tremendous amounts of information to come forth. Information that was significant in other areas of the business, I found, was also vital to the long-range goals of the organization. The open manner in which the information was exchanged brought forth an order seldom found before.

Such a systematic approach brings clearly into focus such areas as needs and wants. It is an opportunity for each participant to give reasons for such considerations. The highlight is when everyone must attempt to quantify the weight given to each item. This continues until there is consensus on the value. Some items may be agreed upon fairly early, while others spark a much more lengthy give and take.

Certainly one of the strongest points of such a process is where potential risks are identified. These risks are ranked as to level or severity and can greatly determine the final decision. Included in this will be the anticipation of possible problems, their causes, and the setting of any alternatives or contingencies. (See Figure 14-3.)

Summary

Organizations cannot function without a capability to solve problems. That's fundamental truth. Determining the optimum method to solve problems is not so basic. The best way, of course, is to prevent problems before they happen. This is done through a combination of data gathering, creation of good operating practices, and long-range planning. This, then, is followed by timely implementation and a continuous effort to maintain control.

If a problem does occur, the manager has to take one or two actions. First, he or she must troubleshoot the problem with an expert capable of coming to the problem rapidly, investigating it, restoring order, and directing personnel in resuming production.

This person may be on staff or acquired through suppliers and consultants. My feeling is that this talent should be on staff whenever possible. This is more likely in a larger company than one with fewer than 500 employees.

Time is inevitably important in problem-solving situations, and it is important to react as quickly as possible to limit liabilities. It is entirely possible to find wide variations in capability of technical personnel or even availability of these services, for any number of reasons. This can often add to the time required for a good resolution.

Where long-term or major decision making is required, it is better to have some form of systematic approach that includes participation of a group of managers from all the disciplines potentially involved. Such an effort brings

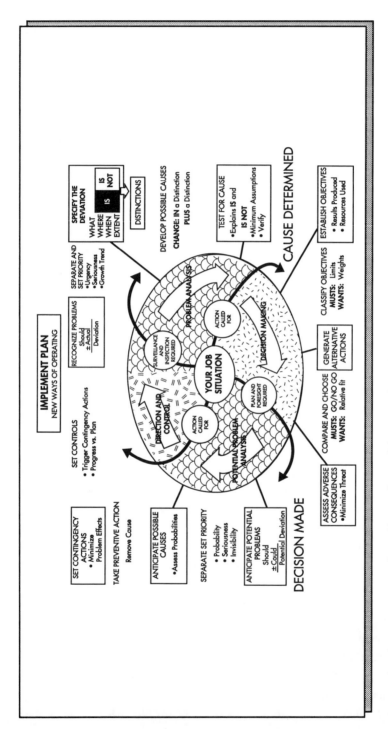

Figure 14-3. *The action sequence for problem analysis and decision making. The sequence runs clockwise systematically around this "wheel," from problem analysis through decision making and potential problem analysis to direction and control. But a manager may enter this sequence at any point on the "wheel," depending upon the demands placed on his job situation. If he clearly knows the cause of a problem he will start with decision-making steps, or if a decision has been made, he will concentrate on potential problem analysis and direction and control.*

more discipline to the procedure and greatly expands the amount of information shared. This type of group activity should be attempted only after formal training in such techniques in order to achieve the levels of quality of decision-making targeted.

Such methods require a change of reasoning from manager to worker. Some individuals do not like to approach decision making in this group format, thinking it dilutes their authority. This failure to admit that such an effort can provide a better decision over an individual one is a problem still prevalent in too many firms.

CHAPTER 15

Maintenance (Not an Expense, a Contributor)

The quality of the paint finish, as well as the quality of the workplace environment, depends largely on the condition and performance of the equipment.

The very subject of maintenance evokes all kinds of thoughts and impressions where finishing and management are concerned. They are treated in so many differing ways and at practically all levels of attention. Maintenance can range from being almost nonexistent to highly professional. It can be a tremendous contributor to quality, employee health, safety, marketing capability, and extending the life of capital investment. It can also jeopardize these.

Many people fail to look at maintenance from the perspective of all these elements. It is normally accepted that paint quality is determined by the level of cleanliness in the system. Dirt has always been the principal quality problem in paint. During my 40 years in the business, I, personally, have been continually vigilant about dirt, and for this reason, many leading companies have made dramatic inroads into resolving the problem of dirt in the finish.

Clean rooms

We are seeing more and more clean-room environments as part of new facilities. Strict rules of dress and conduct go hand-in-hand with these facilities. Investigation has uncovered that such innocent things as workers' deodorant flaking off in minuscule grains into paint caused massive rework. And therein lies the root of the issue. Even microscopic forms of dirt are devastating. A human hair, for example, is about three times the size of most dirt defects coming from other sources in a paint system. Managers need to be cognizant of the many sources of dirt, both human and inanimate, that can destroy a paint shop's performance.

Managers today have to be sensitive to many issues. On the worker side, people want to wear long hair, beards, and mustaches. They prefer muscle shirts and tank tops with no sleeves to shop coats. Even shorts find their way into shops. Managers must diplomatically forbid this attire. We need to keep people's skin shielded from finishing products, not provide greater exposure. Next is the consideration of this dress code on painting quality. Personal dress can greatly increase paint rejects by way of lint and hair particles finding their way onto freshly painted surfaces.

Figure 15-1 lists most of the major sources of dirt found in a typical finishing operation. Awareness of these sources will help in the establishment of maintenance procedures and schedules along with operational guidelines to greatly lower exposure to dirt problems.

In Figure 15-2 is an outline for rules concerning policies used to ensure clean-room performance from the human resource side of our business. Some of these rules are controversial. Smoking and chewing habits, for instance, have to be regulated, and that is difficult in some shops, but finding greater acceptance in recent years. Hair nets in a paint booth is an especially sensitive area, since egos are involved. When you tell some big, burly fellow about 6'4" and 265 pounds, with long hair and a beard that he has to wear a hair net, you will find out quickly how good a diplomat you are.

All of these issues are a part of the ever-changing scene in finishing. Management is required to act on them as part of its effort toward higher first-time yields, lower costs, and regulatory compliance. This then impacts the maintenance and operational sides of the paint operation.

Areas of maintenance concerning air supplies, booth balances, uniforms, eating areas, showers, filters, and respirators impact the health of all associated with finishing in varying levels.

Hazards of paint

The quality of the paint finish, as well as the quality of the workplace environment, depends largely on the condition and performance of the equipment. Many paint formulations are hazardous if inhaled in sufficient quantity. Some paints can cause irritation to the skin. Paint and sludge

Dirt in the Finish

Dirt is one of the most prevalent problems encountered in the finishing industry today because it is the most elusive to solve.

Dirt in the finish may be a result of one or more of the following conditions.

- **The product is dirty.** All loose dirt must be removed from the product prior to entry into the spray area by properly blowing off and tackragging. It is also important to use non-lint-carrying products. If masking operations are performed, tack off the product again after masking to remove any paper, dust, or lint. The additional cost of this operation is made up many times in the reduction of rework.

- **The paint is dirty.** Paints are easily contaminated during mixing, with dirt on containers and mixing equipment the prime suspects. You should wash paint paddles each time they are used. Storing paddles in buckets of solvents often leads to seedy resin agglomerates being formed in the dirty thinner. Some paints may contain fillers that will not mix readily. Lumps, skins, and large particles must be strained out with *metal* or *nylon* strainers. *Do not use rags or cheesecloth* under any conditions with finish materials.

- **Atomizing air is dirty.** Spray guns that are totally immersed in solvent tanks for cleaning quickly become dirt generators. Dirty solvent leaves a residue in the air passageways which will flake off into the air stream when used. Wash the outside of the gun with a solvent rag or brush and wipe dry. Flush the fluid nozzle with a clean solvent compatible with the finish being used. *Remember, do not submerge a paint gun.*

- **Dirty air lines.** Piping and air regulators often become contaminated with rust. Check the condition of your air lines by holding a clean handkerchief or white rag over the air cap with all fluid turned off. Trigger the gun for one minute, causing the air to flow through the handkerchief. Examine the white cloth for signs of rust and/or oil contamination. Always clean the filter at regular intervals to ensure clean air flow through the extractors.

- **Dirty working area.** Good housekeeping in the booth and surrounding areas adds greatly to clean painting. Inspect hoses, booth walls, and conveyor shields regularly to ensure dry overspray is not falling into your jobs. It is your responsibility to be aware of dirty paint hooks and to assist in removing them on a regular basis from the line. Help the maintenance departments do their job by getting these hooks rotated to them. Your maintenance partner in management is a good man to get to know. *Both* of you share in this effort.

- **Improper operation of the booth.** To function properly, a booth depends on a supply of clean air. Air intake filters must be kept clean to ensure this supply. Exhaust filters similarly must be cleaned or changed regularly to ensure proper air movement through the booth. Intake filters are normally more neglected than exhaust filters. All doors on booths and vestibules should be closed at all times to assist in maintaining a balanced booth. If not, exhaust air will be drawn into the booth area from surrounding sanding operations.

Figure 15-1. Dirt can find its way into paint from any of a host of sources.

- **Poor operator techniques.** Excessive air pressures will contribute to dust and overspray accumulation in the spray area. The resulting high air will dislodge other dirt and dust that might not normally fall into a finish. Have a painter trigger the gun to keep slugs of paint from blowing off the fluid nozzle tip. Always check your dry spray on edges of carrier hooks, as dirt here is common. Have your blow and tack personnel blow the hooks before starting to tack off the finished part.

    Too much attention is often directed at trying to make the paint smooth by high or fine atomization. Paint should look a little dry or rough when you first complete your painting. The chemistry of the paint will then permit solvent to reflow the film as it goes into the oven. If you spray too dry, you have no solvent left to assist you in achieving a smooth, deep image paint finish. Remember, a good paint job almost sags.

- **Lint on clothing.** Operators should wear starched, lint-free uniforms or those made of antistatic material, designed for this purpose, such as dacron or nylon.

- **Static electricity.** Dust particles are attracted like a magnet to new paint, especially when the humidity is low. The condition is aggravated by rubbing or wiping, as we must do in our tacking or even sanding activities. Make sure to ground the product, spray gun, and operator when possible. Make sure your hooks are clean and touching a spot on the part. Do not let paint accumulate to heavy thicknesses on these contact points where multiple hanging hooks are used. Many companies use air hoses made of conducting hose material or run an actual ground wire through the air hose. This is not necessary in an electrostatic gun since it is grounded as part of the installation. It is a consideration with conventional paint guns. All fiberglass parts must be blown with deionized antistatic air guns unless grounding is accomplished in other ways, such as through use of conductive primers, etc.

- **Put on enough paint film.** By applying the proper amount of paint on your parts, dirt will not affect you as much. Any particle of dirt or overspray up to 15 microns in size will probably be absorbed into the paint film and resinated if 1.5 mils or more of paint is on the unit. Increase this thickness to 2.0 mils, and you can absorb up to nearly 25 microns in size without detriment to the finish.

    Thin paint films will not do this, and such particles will stand up in the finish. Pigment grinds will often protrude in these films as do metallic particles. If it appears that you have a uniformly dirty painted part, stop the line a moment and put on another coat of paint. Often times this will cure your problems.

    Remember, *you* are the only thing in or around a paint line that has brains and judgment. The paint does not; the parts do not; the equipment does not. You must be the master of these things.

    Like good cooks, who put a little of this and a little of that in their cooking, and then put a whole lot of themselves in the rest of the venture, *you* must put a whole lot of *yourself* in running an area. It is called by names such as skill, pride, or professionalism.

Figure 15-1. (Continued.)

Paint Facility Environmental Policies

- **Dress code.** The goal is to be 100% lint-free. Any aprons, gloves, headgear, and coveralls should be lint-free. The wearing of lint-free clothing is mandatory.
- **Headgear.** Anyone handling paint or working inside paint booths should wear hair nets. Paint caps should be worn in all other work areas on the production floors. Baseball caps are permissible.
- **Eye protection.** Safety glasses should be worn in all areas of the paint facility. Exceptions are locker rooms, restrooms, cafeteria, break area, and offices.
- **Hand protection.** Shall be lint-free. Lint-free gloves are mandatory.
- **Footwear.** Shoe sole wash-off pans should be used.
- **Issuance and return of dress materials.** Non-disposable items can be handled through supply cribs within the department, while disposables can be discarded in containers provided.
- **Smoking/chewing tobacco.** Due to the need for a clean-room environment, no smoking or chewing anywhere in the paint plant should be permitted. The exceptions can be the locker rooms, where employees can receive an air blowoff before entering the clean-room environment.
- **Eating.** Allowed in the cafeteria, locker rooms, and specified break area with picnic tables.
- **Canteens.** Coffee and pop machines can be located in the locker room lobby. All other types of food and drink canteens should be located in the cafeteria. No cigarette machines should be in the paint facility. Coffee and pop machines can be located in the central break area.
- **General housekeeping.** To maintain our clean-room environment, GOOD HOUSEKEEPING is a must. Trash receptacles, "green giants," and balers should be provided.
- **Casual visitor attire.** Visitors should wear smocks and paint caps.
- **Working visitor attire.** Same policy applied as for regular paint plant employees.

Figure 15-2. Strict adherence to clean-room rules will help ensure quality paint.

accumulations aid in the growth of bacteria. Some can be quite harmful, while others can be the source of very unpleasant odors.

When chemicals and overspray from booths are not contained, they cause damage. Such damage can occur outside the plant: Contaminants traveling through the air can damage cars, homes, and other objects. In many cases, heavy payments have been made in the form of rework and fines.

A large amount of overspray in the air can be highly combustible and ignited easily. Paint buildup on walls, floors, and ceilings of booths or any other equipment is flammable and presents a real fire danger. Fire is one of the most feared catastrophes in a paint shop.

One shop, one manager

Good paint managers must on a regular basis see for themselves just what is taking place in operations. Maintenance is often performed on the third shift. At times, the performance of these people is marginal, and often there is a great deal of turnover. The mere presence of a manager in these areas in the middle of the night will not only engender good employee relations, but provide the manager the opportunity to assess training and other needs.

Managers might find in these visits that very little is being done, that money being spent is not yielding value, and, worse, that laxity is creating hazards. I recall a time when I found old paint filters stacked together and smoking in a dumpster. They were about to catch fire from spontaneous combustion, but no one seemed alarmed. The root of the problem was that personnel had changed without the paint department knowing it, because the maintenance department was controlled by a manager outside paint operations.

For this and many other reasons, it is my conviction that executive management should structure finishing operations to be as self-controlling as possible. All the needed activities for the paint shop should be under one manager. Though this structure has not caught on universally, where it has, the organization has enjoyed higher performance and greater productivity. Maintenance of the paint shop and waste treatment works best and is better understood when placed under the people who make the mess in the first place. When those who clean up the mess and get charged with the costs are those making the mess in the first place, greater attention to good housekeeping seems always to prevail.

Another reason for insisting on good maintenance is image. A smart company will ensure that its paint shop is always immaculate. They will use it as a marketing tool to impress the customer. Customers today feel that the level of time and attention paid to finishing a product is a good indication of the integrity of the fabrication and assembly operations of the business.

Managers constantly are asked to take a hard look at their paint shop and ask: If I were a potential client, would I want to place my faith and future in such an organization? Even closer to home, how can I expect people to provide high-quality workmanship in this facility if it is not maintained and functionally viable? If answers come up short, something needs to be changed, and fast.

Organized labor and the organized shop

Another important consideration is the role of organized labor. Managers must deal with issues of negotiated terms that affect just about everything from time accounting to classification integrity. There are issues of premium time, overtime, and work out of classification that managers must deal with.

Often, maintenance is performed on weekends. Work rules governing seniority and qualifications for a "premium" period can make for some resourceful and creative management. This is especially true with today's more sophisticated technology.

There are cases in which people hold a classification in another area of the plant for those times when maintenance is not needed within the paint department. These workers are then eligible to work the overtime in the paint shop for their classification. They often do not know what is required of them in this unfamiliar area. Some will accept work and then decide to not show up, without any penalty to them. At times, they do report, but without proper tools or clothing. Some have health restrictions and allergies not considered when asked to work.

There are electricians who may not know electronics. An equipment plumber or repairman may not know a thing about maintaining robots. There have been cases where people from other areas work everywhere in the plant and trip over limit switches and cause unimaginable havoc. (Case in point: If the line had a computer-controlled conveyor it would show a message that product had passed that point when it had not. It would then shut down the system to avoid a wreck it senses might occur. The cause for shutdown must then be found and overridden before work can begin. The list can go on and on.)

Management, in trying to get around these contractual constraints, may schedule overtime on a daily basis so the primary person on the job is eligible and capable to do the work since it is not a premium day. The piecemeal approach normally does not get the job done well despite the money spent. Management may elect to overpay a person to get them to accept another job so the proper person will then be available for a certain need. This is not efficiency.

I have also observed instances where people brought into a plant to perform work on paint shop maintenance were sent to other parts of the plant. Their timecards, however, were made out and charged to the original assignment and not the one to which they were sent. The accounting system will show that a certain number of hours have been expended on paint maintenance, yet no increase in performance is shown for the following week.

(Other aberrations sometimes emerge when accounting has only partial information. To illustrate, suppose your company has a system in which prime labor is used to develop support labor such as maintenance, but then robots are put into your system. Since less work is being performed by people as prime labor, there is a very good chance the manager of the maintenance department will be told to reduce maintenance...all this at a time when there is a need for more highly trained and sophisticated maintenance. If that maintenance manager is from outside the paint shop, there will be a much heavier clash between the forces involved than if the maintenance were under the paint manager. Once again, a sound reason for self-containment, instead of a series of separate managements affecting paint performance.)

Startup

Another area of maintenance that is often overlooked is the function of starting up facilities each work day. It is easy to forget that this is the other part of maintenance need. We have to get the shop up and running and then keep it running, as well as do the down period maintenance.

The critical days, of course, are Mondays, days after vacations and holidays, and those times following major maintenance deep clean. How many times are stages not heated in pretreatment systems or ovens not turned on and production is rolling through each?

These are the times when a checklist of pre-startup needs is imperative. Such a list will vary from operation to operation, but can still be tailored to your shop's specific needs. An example is shown in Figure 15-3.

The smart manager will also use this chart in reverse as the basis for priorities in creating maintenance schedules. For example, the first item calls for the

PRE-START-UP CHECK LIST
FOR LIQUID PAINT SPRAY LINES

PRETREATMENT/CLEANING
☐ Heaters on and set at proper temperature

☐ Pumps on
☐ Rinse stage overflows operating properly
☐ Solutions at proper concentrations

☐ Bypass discharge valves set at correct pressures

☐ Chemicals available and properly stored
☐ Operator safety equipment provided
☐ Spray nozzles functioning properly

OVENS
☐ Ovens and adjacent areas free of combustibles
☐ Purge timers properly set
☐ Cure oven on and all zones set at proper temperature
☐ Dry-off oven on and set at proper temperature

☐ Exhaust fans operating

PAINT AND PAINT MIX AREA
☐ Ground wires in place and secured
☐ Drum, tanks and other containers properly vented
☐ Exhaust systems operating
☐ Required colors available
☐ Agitators on and adjusted to proper speeds
☐ Paint at proper application viscosity and temperature
☐ Paint pumps on and adjusted properly
☐ Fluid filters pressure drop below maximum

☐ Safety equipment provided

APPLICATION EQUIPMENT
☐ Proper cleaning and maintenance performed
☐ Area free of obstructions

☐ Ground wires in place and secured
☐ Air makeup and exhaust systems operating properly
☐ Fluid lines and applicators purged of wash solvent
☐ Paint flowing to applicators
☐ Applicator positioning, fluid pressures, flow rates, atomizing air pressures and spray patterns correct
☐ Safety equipment provided

CONVEYORS
☐ Free of obstructions entire length
☐ Guards in place
☐ Proper lubrication performed – not excessive
☐ Hooks and fixtures properly hung
☐ Speed set correctly _____

ON-LOAD AREA
☐ Area free of obstructions
☐ Schedule available
☐ Proper hooks and fixtures available
☐ Proper ware staged
☐ Safety equipment provided

OFF-LOAD AREA
☐ Area free of obstructions
☐ Packing instructions available
☐ Packing containers, labels, etc. available
☐ Safety equipment provided

FINAL WALK THROUGH CHECK
☐ Pretreatment/cleaning tanks at proper temperature (actual)
☐ Ovens at proper temperature (actual)
☐ Conveyor at proper speed (actual)
☐ Faulty equipment repaired or replaced
☐ All interlocks and safety equipment functional
☐ Access to emergency shutdown devices clear of obstructions

Association for Finishing Processes of the Society of Manufacturing Engineers
One SME Drive
P.O. Box 930
Dearborn, Michigan 48121
(313) 271-1500 FAX (313) 271-2861

Society of Manufacturing Engineers

Figure 15-3. Pre-startup checks following scheduled and unscheduled shutdowns are essential to averting serious problems caused by oversight.

heaters to be on and set properly. This would then translate into a maintenance item to see that the heaters were serviced on some form of regular basis.

The reverse of this startup procedure is one required for shutdowns, especially for extended periods of time such as vacations or product line changeover. Much damage can be prevented if there is a proper procedure for this. Figure 15-4 is an example of such a method. It may not fit every paint operation, but it points out the various considerations. Use of these guidelines should minimize both facility and safety problems and help ensure the viability of bringing any facility back into production easily.

Pretreatment maintenance

Any good supplier will have a set of suggested procedures for maintenance on pretreatment systems. These schedules will be based upon some combination of metal going through a system, the chemicals being used, and readings taken through daily titrations. These outlines should be obtained and adhered to rigorously. Remember, we do not paint metal in most cases. We paint phosphate as the foundation of the paint work. Dimes spent to ensure the correct performance of a pretreatment system will save dollars further down the way in improved paint quality and reduced field failures. A typical procedure appears in Figure 15-5.

General housekeeping

It is also good practice to develop and adhere to a housekeeping program. Creation of a team to perform weekly inspection of the various elements in finishing operations will bring about a much higher level of awareness to everyone in the discipline. An award could be created to designate either the best or worst performing area. I have seen old crusty-looking paint buckets made into the award for the worst area. It was lighthearted in nature, but it got the point across. People do not like to have such a symbol being placed in a highly visible area of their domain.

Figure 15-6 is an example of a program that can be used to examine and quantify the performance found in these areas of facilities. In this example, simply insert a number to represent each step or level, add the numbers, and divide by the number of booths/ovens reviewed. This will result in an average for all lines. You may want to just score each line as a separate entity and then compare scores against each other if many lines are present.

Suggested Extended Shutdown Procedures

The following considerations will assist in a more orderly and professional shutdown:

1. Determine the type of shutdown and its duration. Is this a long weekend, a vacation shutdown, or a walkout while operating? Order and arrange for all lockers to be vacated for fumigation during these shutdowns. Recover coveralls, uniforms, and company equipment for cleaning.
2. Can you schedule a normal cleanup gap and leave production pieces in the oven? Should you also clear the oven? In either event, see that the parts are baked dry before shutting down any ovens. Will there be maintenance programs going on?
3. See that all spray guns, sanders, blow guns, etc., normally operated with quick-coupler style attachment are cleaned and turned in to the department. They are your assets and responsibility. If mounted permanently, clean thoroughly.
4. Empty all wash tanks in areas and remove scrap paint and scrap solvent from inside buildings. Empty all trash containers and remove from buildings.
5. See that all hose troughs in booths are filled with clean solvent. Keep all quick couplers on paint hoses wet in these troughs to protect seats from drying out and leaking.
6. Blow back, remove, and flush all special paint hoses in use if shutdown is more than 72 hours. Turn off agitators on paint urns and release pot pressure.
7. Remove all flammable production materials from around booths and store in authorized cabinets or paint mix areas.
8. Complete any blended paint batches and wash up all tanks. See that all drums are grounded and fitted with approved spring-loaded spigots or molasses-gate valves. Turn off agitators, but run each color five minutes a day to maintain during long shutdowns over two weeks.
9. All major or junior overhead circulating paint systems must be kept running at all times. TSA enamels should remain stable for a period of four weeks or more without any problem. Examine every few days to verify stability if shutdown is extensive.
10. All chassis or air-dry systems must be monitored closely to prevent oxidation and curing in the paint lines. Underground storage tanks of chassis paint must be circulated and treated with butyl alcohol to retard jelling. See your paint vendor on long shutdowns of one week or longer.
11. Clean up, paint, and repair of facilities, if advised to do so, during long labor strikes. In no other cases should supervision undertake normal bargain unit activities during scheduled shutdowns.
12. Walk all areas of your responsibilities before shutdown, if possible, to verify condition. If labor problems are suspected, assign management to specific areas to ensure an orderly vacating of premises. A safe and orderly posture must be maintained.

Figure 15-4. It is good practice to shut down paint areas during vacation periods, labor negotiations, or other periods of extended non-production.

Suggested Routine Shutdown Procedures

I. **Initial Cleaning and Descaling**
The following procedure is recommended prior to installing chemicals in any stage of a recirculating spray washer system:
1. Drain all chemical solution tanks, making sure that all acid and alkaline solutions are neutralized to conform to local EPA requirements.
2. Inspect tanks and manifolds for leaks and manually remove all monorail hooks and parts that have fallen off during normal production runs.
3. Manually remove all sludge and debris from tanks.
4. Fill tanks with a 5% by volume acidic solution (typically phosphoric acid-based.)
5. Circulate the solution for at least four hours at 75°F to 130°F (24°C to 54°C). If the machine is extremely dirty or cannot be heated, it may be necessary to recirculate the solution for longer than four hours.
6. Drain all tanks.
7. Flush out all tanks with clean, clear water.
8. Fill tanks with clean water and circulate the water for approximately one hour.
9. Drain all tanks.
10. Inspect all manifolds, risers, headers, screens, nozzles, etc., for clogs, cracks, missing components, and general condition. Repair where necessary.
11. Document spray washer condition and repairs made.
12. Fill tanks with recommended solutions of cleaning or pretreatment products.

Admittedly, the above procedure is time consuming. However, the condition of the spray washer is critical to the successful performance of any metal finishing operation. A few hours of cleaning time *before* startup will save many hours of troubleshooting later.

II. **Daily Maintenance Procedure (all stages)**
1. Check and clean pump screens.
2. Clean or replace plugged nozzles.
3. Turn on recirculating pumps, check for misdirected nozzles, and adjust spray pattern. Check specifically for overspray at both the entrance and exit of each stage and adjust as needed.

Figure 15-5. Preventive maintenance procedures for pretreatment systems fosters optimum performance and cost-effective operation.

III. Monthly Maintenance Procedure

1. Drain all solution tanks, making sure all acid and alkaline chemicals are neutralized to conform to local EPA requirements.
2. Inspect tanks and manifolds for leaks and manually remove all hooks and parts that have fallen off during normal production runs.
3. Manually remove all sludge and debris from tanks.
4. Thoroughly flush the tanks with clear water and drain.
5. Fill the tank with clear water and recharge chemicals.

NOTE: The monthly dumping period may be altered based on the production rate, soil loads, etc., in each individual facility.

IV. Six-Month Maintenance Procedure

Industrial spray washers should be cleaned and descaled approximately every six months by following the procedure listed earlier for initial chemical installation.

The time intervals (daily, weekly, etc.) listed in this report may vary from spray washer to spray washer. They are based on average job conditions and do not consider variables such as heavily soiled stock, high production requirements, etc., when determining the frequency of cleaning the equipment.

V. Repair Immediately

1. *Pumps.* Leaking pumps should be repaired as soon as possible in order to ensure adequate spray pressure.
2. *Controls.* Spray pressure and temperature controls must be maintained in good operating condition to give the operator an effective means of controlling these parameters.
3. *Valves.* All makeup water and drain valves must be functioning properly at all times to ensure accurate solution control.
4. *Gages.* All gages vital to the proper operation of the unit should be operational and **accurate**.

Figure 15-5. (Continued.)

PAINT BOOTH — OVEN MAINTENANCE

PLANT _____

DEPARTMENT _____

DATE _____

BOOTH NAME _____

ITEM DESCRIPTION	1 EXCELLENT	2 GOOD	3 SATISFACTORY (NORMAL)	4 FAIR	5 UNSATISFACTORY
GRATES					
ELIMINATORS					
DECK PLATES					
PLASTIC COAT					
BOOTH SWEPT					
BOOTH WASHED					
TRACK SWEPT					
TRACK STRIPPED					
PIT					
SANITARY SHIELD					
BACKWALL					
SPRAY NOZZLES					
SKIMMED					
LIGHTS CLEANED					
LIGHT GLASS BROKEN					
LIGHTS BURNED OUT					
VESTIBULES					
FANS					
STACKS					
STANDS					
HOSES					
FILTERS					
TOTAL					

Total Accum Rating _____ ÷ Facilities reviewed _____ = Avg Rating _____

REMARKS

EVALUATED BY: _____

_____
COLUMN, BUILDING, DEPT., ETC.

Figure 15-6. Housekeeping checklists such as this reduce the potential for misunderstanding of tasks needing to be done.

Figure 15-7 can serve as a guide for reviewing other areas of the finishing operation outside the booths and ovens. Once again, this permits the creation of a numerical score in judging performance.

It is important to move people in and out of the team every quarter or so. By so doing, more people can participate and become a part of the awareness effort from both sides.

Over the years, it has been my good fortune to have put together some additional tools to assist in identifying problems associated with spray booths, ovens, and compressed air supplies. These have also been put into chart form, Figures 15-8 through 15-10, for easier use for both maintenance and production personnel. They have been provided for your use.

No discussion on maintenance would be complete without comment on some of the new trends currently emerging. These are changing the way in which traditional activities are performed.

A much more demanding and sophisticated customer has forced industry to change. Simple needs that were satisfactory years ago no longer keep the customer from looking toward the competition. Today, they carefully assess a product's reliability, quality, and performance. Industry must both identify and address these continually evolving needs if it is to compete effectively in a world marketplace.

This means paint shops must critically and regularly evaluate their methods of operation. This must include not only the areas mentioned, but must take into account such things as waste disposal and water treatment needs. These join with all the others in generating the best performance while operating within ever more stringent regulations.

Wastewater

The areas of wastewater and sludge collection were non-issues 30 years ago, but to paint managers today are major areas of concern as they evaluate new approaches to total control of their operations. Wastewater treatment is a particularly sensitive area and one in which managers tend to make mistakes. Figure 15-11 lists the 10 most common of these mistakes as viewed by Richard Dusley. In many production facilities, the paint areas produce most of the industrial waste. Because of this, executive management in many companies is beginning to place these activities under paint managers who know something about the waste, instead of more traditional managers.

HOUSEKEEPING

PLANT _____

DEPARTMENT _____

DATE _____

ITEM DESCRIPTION	0	1	2	3	4	5	6
MACHINE TOOLS CLEAN & PAINTED (2 YEAR CYCLE FOR PAINTING)							
TOOLS & FIXTURES PROPERLY STORED							
WORK AREA CLEAN & ORDERLY (DIRT, OIL, GREASE, DEBRIS)							
FLOOR CLEAN & IN GOOD CONDITION (LOOSE BLOCKS, ETC.)							
AISLES CLEAN & ORDERLY & IN GOOD CONDITION (NOTHING OVER AISLE LINES IN AISLE)							
AISLES PROPERLY IDENTIFIED – LINES							
TRASH BARRELS CLEAN, PAINTED AND IN PROPER LOCATION							
SAFETY ITEMS IN THEIR PROPER PLACE							
FIRE PROTECTION EQUIPMENT IN DESIGNATED LOCATION							
MATERIAL STORAGE NEAT & ORDERLY							
FORK LIFTS, WORKSAVERS, JACKSTACKERS PAINTED & IN GOOD CONDITION							
WINDOW CLEAN & IN GOOD REPAIR							
OUTSIDE STORAGE AREAS CLEAN & ORDERLY							
ROADWAYS CLEAN, ORDERLY & IN GOOD REPAIR							
ALL AREAS FREE OF UNNECESSARY PAINTINGS & POSTINGS							
RESTROOMS CLEAN & ORDERLY (CHECK DAMAGED FACILITIES, SOAP, TOWELS, PAINT, ETC.							
CANTEEN AREA CLEAN (CHECK CONDIMENTS, MACHINES, ETC.)							
OFFICE CLEAN & ORDERLY (PAPERS & RECORDS STORED, CONDITION OF FURNISHINGS & PAINT, ETC.)							
TOTALS							

KEY
% Points

100 — 6
90 — 5
80 — 4
70 — 3
60 — 2
50 — 1
40 — 0

General Comments: Review should include necessity of improved painting and replacement of damaged facilities. Inspector should take into consideration location and type of use of facilities.

Rating Scale:

	Unsatisfactory		Competent		Distinguished
%	60	70	80	90	100
		Provisional		Outstanding	

Total Accum Rating _____ ÷ Facilities reviewed _____ = Avg Rating _____

REMARKS

EVALUATED BY: _____

_____
COLUMN, BUILDING, DEPT., ETC.

Figure 15-7. Schedules for general housekeeping help ensure that all tasks are completed on time, every time.

Maintenance (Not an Expense, a Contributor)

SPRAY BOOTH

FAULT	RESULT	Dirty Job	Thin Coats	Poor Opacity	Sags	Overloading	Popping	Softness	Overspray	Uneven Application	Recoat Failure	Fire Hazard	Water Splashes
Dirty filters	Vacuum in booth (hot air drown from oven)	*		*		***	***	***	*	*	***		
	OR												
	Not pressurized and dirty air drown in from prep. area)	***	*** (Poor build)										
Breached or damaged filter	Turbulence	***			**				**	**	x		
	Over-pressurized				*** (in oven)				**	*	x	***	**
Water level Low	Increased extraction		***			***	***	(Cold air forced into oven)	**	**	***		
High	Restricted extraction		***	*	**				*	*	***		
Empty	Increased extraction with buildup of dry paint in reservoir	***										***	*** (Alkaline contamination)
Use of incorrect water additive, or incorrect use of water additive	Blocked water jets and filters. Formation of dry powder on antisplash panels +Bacterial/germ cultivation (unpleasant smell) Corrosion of plant	***					NOTE: 1. Use only lint free overalls in the spraybooth (use them only for this purpose). 2. Clean and blow off prepared vehicles outside the spraybooth paying particular attention to the engine compartment. Do not exceed 40 psi. 3. Repair damaged or ill-fitting spraybooth doors promptly. 4. Affix clean sheet of paper daily to spraybooth wall for testing gun.				**	***
	Paint deposits difficult to remove												
Matting paper, rags, masking paper, old cans, etc., in booth	Dirt accumulation	***										***	
Spraying on walls in booth	Poor light reflection									*	x		
Loose deposits of dirt, dry spray, rust, etc., on booth walls	Dirt in atmosphere	***											
CODE: *** Most likely failure to be associated with the fault. ** Likely failure. * Failure less likely to be associated with the fault. + Health hazard. X Will affect color of metallics.													

Figure 15-8. Spray booth troubleshooting matrix.

OVEN

FAULT	RESULT	Popping	Softness	Dirty Job	Overspray	Impaired Durability	Polishing Impaired	Fire and Explosion Hazard	Loss of Gloss	Recoat Failure	Discoloration
Dirty Filters	Diminished air velocity	*** (Upper parts)	*** (Lower parts)			*	**			***	
	Diminished oven pressure		*** (Cold air drawn from booth)		*** (Drawn from spraybooth)	*	**				
	Spraybooth/oven pressure inbalance			***							
Filters damaged or breached	High velocity jet streams and turbulence	*** (Local)	** (Local)	***		*					
Thermostat Probe Not correctly sited in moving airstream. And/or insufficiently sensitive	Excessive high/low temperature modulation	***	***			*	*				
10% bleed duct closed/10% makeup filter clogged	Foul oven Excessive fumes					**		***		*** (Microshrivel)		*** (Chemical reaction)
Failure to remove deposits of rust, dust, and flaking paint from oven surfaces	Excessive dirt circulation			***							
Failure to clean unpainted areas on vehicles. Failure to clean masking or remask. Operators entering oven with dirty overalls	Unnecessary dirt introduced into oven			***							

NOTE: Repair ill fitting or damaged oven doors immediately

CODE: *** Most likely failure to be associated with the fault. ** Likely failure. * Failure less likely to be associated with the fault.

Figure 15-9. Troubleshooting the paint oven.

Maintenance (Not an Expense, a Contributor)

COMPRESSED AIR SUPPLY

FAULT	RESULT	Blistering	Non-drying	Poor Adhesion	Contamination	Poor Atomisation	Poor Flow	Over-loading	Sags	Popping	Slow Application	Off-shade Metallic	Uneven Application	Dry Spray	REMEDY
Oil/water not adequately condensed out	Oil/water at spray gun	***	*	***	*										Ensure regular drainage of air receiver, separator and transformer. Site transformers of adequate capacity in cool places. Lubricate compressor with recommended grade of mineral oil of good emulsifying properties
Long air line. Inadequate internal bore of air line. Fittings, compressor, air transformers, and regulators of inadequate capacity	Pressure drop					**	*	**	**	**	*	**			Ensure adequate air supply with 30' 5/16" internal bore air line with appropriate fittings. NOTE: Reduction of viscosity to give improvement may produce other defects
Inadequate compressor capacity. No pressure regulator, regulator diaphragm broken	Pressure fluctuation							**	**	**		**	**	**	
Compressed air intake filter breached. Transformer filter not properly maintained. Compressor sited in dusty area	Dirt in compressed air														Dirt
														***	

CODE: *** Most likely failure to be associated with the fault.   ** Likely failure.   * Failure less likely to be associated with the fault.

Figure 15-10. Locating failures in the compressed air supply.

195

The 10 Most Common Mistakes in Wastewater Treatment

1. Inadequate pretreatment
    - Not removing the bulk of the oil
    - Not segregating certain systems
2. Collection/Equalization Tank Capacity Too Small
    - Can lead to highly variable wastewater which is difficult to treat
    - Can lead to surges of water through system
        - Disruption of sludge bed
        - Not enough time for processes to work
3. Incomplete Removal of Free (Non-emulsified) Oil
    - Problem with pH controller or acid feed
    - Skimmer not working properly
4. Coagulation Step is Incomplete
    - Insufficient coagulant to destabilize colloids to form "pin floc"
    - Too much coagulant fed
    - Inadequate mixing
5. Metal Hydroxide Formation is NOT Complete Before Addition of Flocculant Chemical
    - Feeding incorrect amount of lime or caustic
    - Incomplete reaction — tanks too small or flows too high
    - pH control problems — failure to adequately calibrate, clean, and maintain monitoring and control equipment
6. Floc Not Forming Properly for Good Separation in Clarifier
    - Problem with mixing, feed rate, or application point of flocculant chemical
    - Improper flows or mixing of tank as floc is forming
7. Not Regulating Chemical Feed Rates Based On Wastewater Flow Rates and System Demands
8. Generating Sludge That Contains Too Much Water — Resulting in High Disposal Costs
    - Too much alum being fed — sludge retains water
    - Polymer overfeed — filter blinding
    - Insufficient sludge "aging"
9. Trying to Use a Mechanical Solution for a Chemical Problem
10. Wastewater Flow Rates Through System are Too High
    - System is too small for the volume of water being processed — could be intermittent or continuous
    - Could apply to entire system or just one process

Figure 15-11. Twenty-first-century paint managers must be acutely sensitive to wastewater and sludge and the hazards inherent in treating them.

Some managers, in attacking the wastewater problem, are entering into a form of partnership with suppliers. Increasingly paint companies are acquiring capabilities to supply nearly all the needs of a production paint operation. They have pretreatment chemicals, drawing compounds, booth maintenance supplies, and wastewater treatment chemicals. Not only do some of them supply the products, a few have even entered the field of management services. It is possible today to obtain both production and maintenance management from outside resources.

These partnerships have led to formation of others as well. They can range from total outsourcing of the management of the paint shop down to informal arrangements based on mutual respect and understanding. The level of "partnership" is generally based upon the variety of needs of each of the involved companies.

In a number of operations, the maintenance of paint systems has been put on a contracted basis with the production management still maintaining overall control of the operation. Still other situations will find a supplier provide a full or part-time technical person to the operation to do specific functions. This is usually a free service, depending on the volume of product purchased for that activity.

The important change is that suppliers today are much more than their name implies. A proverbial smorgasbord of services is available today from suppliers and they stand as a valuable resource to the wise manager who would approach doing business differently.

Probably the most effective tool devised in recent times is SPC, a disciplined approach to data gathering. SPC data on equipment is used to predict potential problems and expected life cycles. Its rising importance is presented graphically in Figure 15-12.

Preventive maintenance

The other leg of the maintenance business is one of preventive efforts. No operation can exist and perform well if it just waits until something happens. As we stated earlier, the best gallon of paint or pound of chemical is the one you don't have to use to get the job done properly. This can be adapted to maintenance as well. The best way to avoid a downtime problem is to not let it happen in the first place. That is what preventive maintenance is all about.

Many paint operations today are becoming quite sophisticated in logging data and turning that information into solid programs of preventive mainte-

Figure 15-12. Statistical process control is critical to predictive maintenance.

nance. Such programs let operators know when a bearing is running hot or if water flows are changing in a paint booth water wash system. The systems can predict the cycles of life for paint filters and numerous other replacement parts.

Just as we need a program to ensure that systems are all turned on and adjusted each working shift, the manager needs a good preventive program to help ensure smooth operation and take most of the surprises out of the facilities end of the business.

Collaboration

Managers tend to be territorial in nature. They act almost like the animal world in many ways, spending a great deal of time going around the boundaries of their territory, marking it each day. It can be done in a dozen different ways, but it is done. This is either to reinforce the manager's sense of turf or let potential rivals be made aware of their crossing into foreign territory. Sharing or changing the approach to getting things done, to them, has the effect of undermining their authority or stature. Real-world experience has proven this wrong, of course, but it is difficult to change many managers, especially those who have been in the habit of practicing it for many years.

When managers realize they are not losing their authority, but adding to their professional arsenal of tools, more will begin creating ways to bring the suppliers into their organizations as partners. It is still management, but in a different form.

To further illustrate, cross your arms in front of your chest in "X" fashion. Then extend your first finger on each hand straight out. This is the traditional salute given by those associated with a paint problem in the past, meaning it was always someone else's fault. When people are part of a partnership relation, the salute disappears. It is not everyone else, it is now us.

Summary

To be sure, maintenance is an essential element in any organization. It makes the difference between operating at low or high levels of performance. Maintenance can provide a great deal more than the normal aspects of operational considerations. It is a major contributor to employee health and safety, facility safety against fire/explosion, and the capital investment life of the facility itself.

Proper maintenance can create an image of product quality and enhance marketing efforts to retain customers and add new ones. The paint shop should be the one showplace management can always count on to show a customer without prior notice. If this cannot be done, then your paint operation is not being maintained properly or managed well, both of which lead to degraded performance.

One of the first responsibilities of a maintenance organization is to see that all the facilities are up and running. A startup checklist geared to your operations is a valuable tool to help guarantee that this happens each shift. This list can also be used to determine many items of preventive maintenance. There must also be a shutdown procedure for short-term (daily) and longer periods.

Maintenance today has taken on greater significance and involves many more potential opportunities for management than in the past. It is possible to obtain outside maintenance specialists from companies. This permits a value analysis to be made to determine both needs and dollars required. These then could be compared with costs and timing of having the maintenance done internally.

The objective of such an evaluation would be to improve the internal performance rather than outsource the problem. Experience teaches that such an action for short-term benefits tends to come back to haunt operations in various ways. At best, there is a lessening of loyalty when people see work being contracted out. At worst, there may be cases of sabotage and poor performance over longer periods of time. Myriad examples exist of both success and failure where the outsourcing of maintenance work has been performed. Each case has to be researched and reviewed on an individual basis to determine why one worked and another did not. My feeling is that management should look to its own people first before eliminating jobs.

A better choice in most instances is to create some level of partnership with suppliers after a thorough examination and competitive comparison of what goods and services can be brought to your organization in a value-related approach. This sharing of mutual efforts tends to be a win-win situation.

The creation and use of SPC methods and predictive approaches to maintenance is probably the latest and best way to improve overall performance. The best way to improve maintenance problems is to avoid them.

Using data to formulate preventive maintenance programs should eliminate downtime and lost production. It should cause a higher yield and much higher return for the labor and materials used, as opposed to waiting for a breakdown and then having to use reactive style management.

Maintenance can no longer be viewed merely as a cost to an organization. The competitiveness of the market will not permit it. A good manager must see that maintenance properly conducted is structured so the dollars spent can be traced to a degree where they are easily justified as a contributing part of the paint operation. This contribution must include not only the operational portion of activities, but those of other areas such as health, safety, marketing, and regulatory compliance. In the past, such supporting issues have not been quantified and made a part of the total cost/value yield equation.

Maintenance when performed correctly is not an expense. It is instead a major contributor to improved performance and profitability.

CHAPTER 16

Environmental, Health, and Safety Considerations

The best gallon of paint or pound of chemical is the one we don't have to use to get the job done.

Volatile Organic Compound (VOC) has found its way into the everyday vocabularies of virtually everyone in manufacturing, especially in the painting industry, where VOCs are found in the solvents used in paints and for paint cleanup. But VOCs may also include the materials that evaporate from resins in a curing oven. Paint managers must have intimate knowledge of the materials in their shops to stay within the dictates of ever-tightening restrictions. (See Figure 16-1.) To assist you in working with the myriad regulations dealing with these materials, some definitions are in order.

Compound. A distinct chemical which contains two or more atoms (the smallest unique unit in chemistry).

Density. Weight per volume of a material, usually listed as pounds per gallon or grams per liter. VOC regulations often discuss solvent or VOC density (the density of the volatile material only), coating density (the density of the whole paint or coating), and solvent content (the amount of solvent in the whole paint which is different than solvent density). The generic solvent according to the U.S. Environmental Protection Agency (EPA) has a density of 7.36 lb/gal.

Exempt solvents. Specific solvents identified in federal or state rules which are not subject to certain emissions limitations. Some regulations

> **Environmental Regulations Affecting Painting Operations**
>
> Clean Air Act
> - Ozone nonattainment
> - Air toxics
> - Permits
> - Ozone depleting chemicals
> - Enforcement
> - Formulation versus control equipment

Figure 16-1. New and emerging governmental regulations mandate that paint managers become more knowledgeable of the chemical makeup of materials under their purview.

have exempted methane, ethane, 1,1,1-richloroethane (methyl chloroform), methylene chloride, and certain chlorofluorocarbons.

Organic. Chemicals that contain carbon, with the exception of carbon dioxide, carbon monoxide, carbonic acid, metallic carbides or carbonates, and ammonium carbonate. The most common chemical joined to the carbon atom is hydrogen.

Photochemically reactive. A chemical or material that undergoes changes when exposed to light, typically sunlight. The former liquid carbonics rules defined these materials by the percentages of various constituents which contributed readily to urban smog formation.

Specific gravity. A measure of density. Water has a specific gravity of 1 and has a density of 8.34 lb/gal or 1000 g/l.

Vapor pressure. A measure of how quickly a material will evaporate at a set temperature and pressure. The higher the vapor pressure, the faster it evaporates (volatilizes). Some rules will define a VOC by its vapor pressure.

Volatile. Evaporating readily at normal temperatures and pressures; those temperatures and pressures which define a material as volatile may be different in different regulations. Warmer temperatures and lower pressures will increase evaporation rates.

Solvents are frequently divided into various categories such as chlorinates (methylene chloride), aliphatics (hexane and mineral spirits), aromatics (toluene and xylene), alcohols (methanol), amines (isopropyl amine), glycols (propylene glycol), glycol ethers, esters (ethyl acetate), ketones (MEK and acetone) and others (2-nitropropane). Solvent suppliers often provide a solvent properties chart which lists the density, specific gravity, boiling point, flash point, evaporation rate, etc. This information may also be found in numerous handbooks.

Solvents are often called by different names, which can lead to confusion when attempting to determine which rules may apply to your materials. For instance, methyl ethyl ketone is 2-butanone or MEK. Often the Chemical Abstract Service Number (CAS or CASN) is listed as a cross reference. Since each solvent has only one CASN, it can be used to check whether or not different names refer to the same solvent.

The concern directed at solvents arises from their potential detrimental effects on human safety and health and the environment. Some general concerns are listed below.

- Many solvents are flammable and form explosive vapors; therefore, storage, handling, transportation and disposal of these materials is a safety concern.
- The inhalation of solvent fumes has been linked to various adverse health effects dependent upon the solvent(s) inhaled, the amount, and the length of time involved. Many solvents are central nervous system depressants such as the commonly consumed ethanol found in beer, wine, and liquor. Likewise, solvents can damage certain organs such as the liver and kidneys. Some solvents have been identified as possible human carcinogens.
- Another common symptom of overexposure to solvents is dermatitis. This condition is caused by defatting of the skin and leads to red, dry skin, cracks or fissures, and sensitivity.
- Because small amounts of solvents can cause harmful effects in humans, drinking water contamination is of primary concern. Sources of drinking water — municipal treatment works using rivers or wells and private wells — may be contaminated from leaking storage tanks, transportation accidents, wastewater discharges, or leaking waste disposal sites.

Most solvents react in warm, sunny weather to form smog. Smog is created when sunlight changes nitrogen oxides (NOx) and solvents into ozone (O_3), a reactive and corrosive form of oxygen, as shown in Figure 16-2. Ozone is irritating to the eyes and lungs, especially to sensitive groups such as infants, the elderly, and those with pre-existing respiratory ailments. Ozone also damages trees and commercially grown vegetables, as evidenced by yellowing leaves. Increased corrosion on buildings and statues has also been attributed to ozone, Figure 16-3.

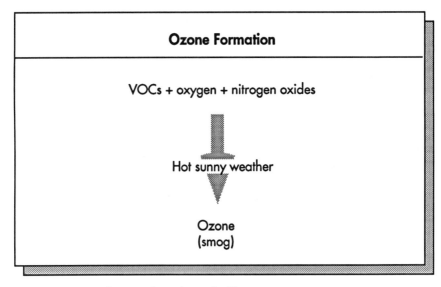

Figure 16-2. Paint solvents are hazardous to health in numerous ways.

Air Pollution
Solvents Create Ozone

Ozone = O_3, a reactive form of oxygen

Ozone in the upper atmosphere protects us from ultraviolet radiation. This layer is destroyed by freons and does not mix with the air we breathe.

Ozone in the air we breathe causes irritation to the eyes and lungs and damages plants, metal, and concrete.

Figure 16-3. Ozone, benign in the atmosphere, is dangerous and destructive at ground level.

Ozone depletion involves the same chemical, but in a different place. Ozone in the lower atmosphere where we live and breathe is not good. However, the same reactive properties of ozone that make it harmful here, make it beneficial in the upper atmosphere. Unfortunately, the ozone in the lower atmosphere does not last long enough to make it up to the upper regions. The ozone already up there prevents ultraviolet (UV) rays from reaching us where they cause skin cancer. Freons and 1,1,1-trichloroethane are being phased out because these chemicals are able to migrate into the upper atmosphere and deplete ozone before it reacts with UV rays.

Although we are primarily concerned with the air pollution regulations for solvents, or VOCs, we would be remiss if we didn't briefly discuss the other environmental regulations which affect users of solvents. Three Federal agencies have enacted laws which greatly impact the use of solvents. The Environmental Protection Agency has rules governing the amount of solvents that can be emitted into the air, discharged in wastewater, and contained in drinking water. Routine uses and accidental releases of certain solvents must be identified in community right-to-know reports. Specific handling and disposal requirements must be complied with when getting rid of unused or waste solvent-containing materials.

In addition, the Occupational Safety and Health Administration (OSHA) sets permissible exposure limits (PELs) for the inhalation of many solvents. They also require employee training, material safety data sheets (MSDS), and safety storage of flammable materials.

Finally, to close the loop, the Department of Transportation regulates the shipment of hazardous materials and hazardous wastes.

The dilemma facing paint managers is that just when they feel comfortable with the U.S. Federal laws, the states publish slightly different or more stringent versions of each Federal rule. Add to that, county and city governments who also write environmental regulations. The overlap is mind-boggling (see Figure 16-4). Each authority may have different rules, forms, fees, and fines.

Then, if your company has locations in Canada, Mexico, or overseas, you will be subject to even more variations in rules. In this chapter we will focus on the U.S. Federal requirements and, as appropriate, touch state and Canadian rules.

Hazardous wastes

The Resource Conservation and Recovery Act (RCRA) regulates solid waste disposal. Creative definitions, though, do not exclude liquid wastes other than some selected wastes such as wastewater treatment plant effluents. The hazardous waste regulations for generators (those that produce the waste) are those of greatest concern to the painting industry. However, the generator has liability from "cradle to grave," so familiarity with transporter and treatment/

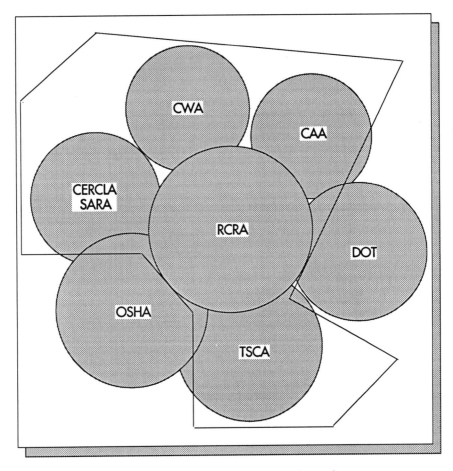

Figure 16-4. Overlap of environmental agencies and the regulations they generate present a peculiar challenge to today's paint manager.

storage/disposal facility (TSDF) rules is beneficial. The hazardous waste rules are found in RCRA document 40 CFR, Parts 260-265, Figure 16-5.

A waste is a *regulated* hazardous waste if it is characteristically hazardous or listed hazardous. Listings in the regulation include specific processes (which usually relate to chemical manufacturers) and specific solvents. Typical examples include F001 and F002 wastes (methylene chloride, 1,1,1-trichloroethane), F003 wastes (xylene, acetone, MIBK), and F005 wastes (toluene, MEK, 2-nitropropane). Characteristic hazardous wastes are either ignitable (D001), corrosive (D002), reactive (D003), or toxic (D004 and up) as

> **Environmental Regulations Affecting Painting Operations**
>
> Clean Water Act
>
> • Direct discharge (NPDES)
> • Indirect discharge (MPP)
> • Storm water - toxics

Figure 16-5. Treatment and transport of waste are now integral to paint shop planning, budget, and operations.

defined in the regulations. Toxics are those which fail to pass extraction procedure tests for specific metals, solvents, and pesticides.

The wastes most likely to be hazardous are discarded paints and solvents (listed, ignitable, toxic), empty containers with too much residue to be legally empty (listed, ignitable, toxic), paint sludges (toxic), and miscellaneous materials such as filters, masking, etc. One special circumstance involving listed solvents which should be avoided is the spraying of color change waste into waterwash. Because a listed waste is hazardous, no matter how you dilute it, all of the water and sludge in the pit becomes hazardous when you put purge solvent into it. However, the same solvents in oversprayed paint will not make it a hazardous waste. (See Figure 16-6.)

Wastewater

The Clean Water Act (CWA) regulations are found in 40 CFR Part-122. The CWA regulates wastewater discharges to publicly owned treatment works (POTW) and to oceans, streams, lakes, and other waterways. Regulations passed in 1993 expand the authority of the EPA to limit pollutant discharges from storm water runoff in addition to process wastewater.

The portions affecting solvents or VOCs are related to discharges of toxic substances listed in Table 16-00 of the regulations. These include several solvents. Since solvents are organic materials, they provide food for microor-

Environmental Regulations Affecting Painting Operations

Hazardous and Solid Wastes (RCRA)

- Sludge disposal
- Paint/solvent disposal
- Filters, masking, etc.
- Purge cycles into water
- Listed wastes/characteristic wastes

Figure 16-6. Discarded paints, solvents, containers, sludges, and various materials require special handling under increasingly stringent regulations.

ganisms and create a Biochemical Oxygen Demand (BOD) and a Chemical Oxygen Demand (COD). One or both of these may be limited in a facility discharge permit to the POTW or streams. In addition, spill control plans must be developed for storage of hazardous substances (Figure 16-7).

Right-to-know laws

The Emergency Planning and Community Right-to-Know Act (EPCRA) had numerous names. It was enacted under the Superfund Amendments and Reauthorization Act (SARA), which reauthorized and expanded the Comprehensive Environmental Response, Compensation and Liability Act (CERCLA). However, EPCRA, or more commonly SARA III, is a standalone act.

SARA Title III, in 40 CFR—, has four major elements, all of which affect users of solvents. It requires emergency planning notification (311 reports), emergency release notification (spills), chemical inventory reports (312 reports), and toxic release inventory reports (313 reports). Participation on the Local Emergency Planning Commission (LEPC) may also be required.

The other right-to-know law is found in 29 CFR 1910 (Figure 16-8), which is an OSHA regulation. This is the Employee Right-to-Know Law or Hazard Communication Standard. Hazard Communication requires employers to maintain MSDS and train employees about chemical hazards. A complemen-

VOCs Affect Your...

Drinking water
(Contamination of streams and groundwater)

Health
(Ozone, air toxics, and occupational exposures)

Plants
(Ozone)

Figure 16-7. The latest (1993) regulations add limitations of pollutant discharges from storm water and provide rules governing discharges to water treatment plants and waterways.

Environmental Regulations Affecting Painting Operations

Toxic Chemical Inventory and Releases

- Annual 312/313 reports
    - LEPC participation
    - CERCLA emergency notification
    - Waste minimization reporting

Employee Right to Know (OSHA)

Figure 16-8. Provision for notification of community and work force of the presence of hazardous chemicals is now mandatory.

tary regulation for laboratory workers is found in 29 CFR 1910.1450. Permissible exposure limits for occupational exposures are listed in 29 CFR 1910.1000.

Air pollution regulations

The air pollution control regulations are promulgated by the EPA under the authority of the Clean Air Act (CAA). In 1990, sweeping revisions were made which greatly impact the coating industry.

The current air pollution regulations can be divided into stationary source standards for businesses, mobile source standards for vehicles, and ambient standards which establish health-based levels of pollutants in the air. The regulations most likely to affect the painting industry are found in 40 CFR Parts 52, 60, and 61.

The main impact of the existing CAA regulations on the coating industry is the result of reducing the emissions of VOCs that contribute to ozone formation. Except for EPA-issued rules for certain large new sources (New Source Performance Standards or NSPS) in specific industry groups such as automobiles, plastic business machines, etc., most rules are issued by the states.

Typically, the state rules will be based upon federal policies such as the Control Technology Guidelines (CTG), which define Reasonably Available Control Technologies (RACT) for creation industry groups. These RACT guidelines require coating formulation restrictions or add-on VOC controls. RACT categories include automobile and light duty trucks, miscellaneous metal parts, metal furniture, coil coating, can coating, etc.

Usually, the formulation restrictions will limit the amount of VOC in a wet gallon (excluding water) or the amount of VOC emitted in the application of one gallon of paint solids to the part. Both of these calculations are discussed in greater detail later. The amount of VOC is not typically an absolute limit. Rather, the limit is a daily volume weighted average of all coatings employed so that the use of low VOC coatings will offset the impact of high VOC coatings. RACT rules may also set technology standards such as minimum VOC control efficiencies and transfer efficiencies.

The RACT rules were written to be the limits for major sources existing at the time rules were promulgated. New major sources of VOCs must, at a minimum, meet the RACT limits. However, a major source may be in the eyes of the permit writer. Usually the emissions used to determine applicability are the potential emissions at maximum operating capacity, 24 hours per day.

Furthermore, many states have applied the RACT limits to new or existing facilities which are not major sources of VOC.

In addition to the RACT rules based upon the CTGs, many states have rules which regulate VOC emissions from pollution sources not specifically addressed in these federal guidelines. These non-RACT rules may be similar in structure to the RACT rules or they may resemble the old liquid carbonics rule, also known as Rule 66. These rules limit hourly and daily emissions rates based upon the designation of the solvent mixtures as photochemically reactive or nonphotochemically reactive and whether or not the coating is heat cured.

As an example, the regulation which sets limits for most of the Navistar Truck operations in Ohio is the miscellaneous metal parts and products coating rule of the Ohio Administrative Code (OAC). Specifically, OAC Rule 3745-21-09(U) limits the emissions from all lines which apply a coating to metal products not otherwise regulated or exempted by the OAC. Navistar must demonstrate a daily volume-weighted average of 3.5 pounds of VOC per gallon of coating, excluding water, for their extreme performance coatings. Other coatings with separate limits in this paragraph are clear coatings (4.3 lb/gal), zinc-rich primer (4.0 lb/gal), etc.

Proper attention to definitions in the rules is needed to determine the applicability of any rule to your facility. For instance, a "coating" in the Ohio rules includes paints, varnishes, sealers, adhesives, inks, etc. A "coating line" does not require that the work be performed in conjunction with a booth, dip tank, oven, flash-off area, or any other device.

The Clean Air Act Amendments (CAAA) of 1990, Figure 16-9, have introduced many new aspects to the existing air pollution regulations. The EPA will be issuing new regulations at an unprecedented pace over the next few years to meet the statutory guidelines established by Congress. A brief synopsis of several important new features of these upcoming regulations is included here.

Areas which are designated ozone nonattainment must show a 15% reduction in VOC emissions in the next six years. Thereafter, if the area is still nonattainment, it must reduce the VOC emissions inventory by 3% annually. The stationary sources will once again be hard hit by new or tightened regulations attempting to meet these quotas. (See Figure 16-10.)

Title III of the CAAA addresses toxic air emissions. A list of 189 chemicals will be regulated over the next decade to protect human health, especially by preventing cancer. Rules will be written for industry categories. The EPA has

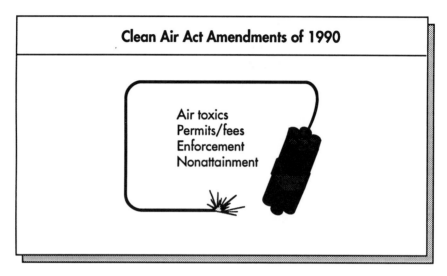

Figure 16-9. Broad revisions to the U.S. Clean Air Act affect the coating industry by reducing the level of acceptable VOC emissions.

already published the source category list. Example categories include surface coating operations general solvent uses, large appliance coating, plastic part coating, large aircraft coating, and flat wood products. Within each category, specific pollutants will be regulated based upon their use or point of emission (e.g. coating operation, coating mixing, equipment cleanup). The coating industry will not only have to meet these air toxic regulations, but also those for categories such as space heaters, boilers, process heaters, wastewater treatment systems, cold cleaning, and cooling towers. A list of some of the organic chemicals regulated by the CAAA is shown in Figure 16-11.

The CAAA requires that sources emitting more than ten tons per year of air toxics demonstrate the use of Maximum Achievable Control Technology (MACT). MACT will be equal to the average control currently achieved by the top few percent of the sources in that category. As technology improves, the level of control achieved by the top controlled sources will increase, and so will the MACT standard.

For the first time, the EPA will require operating permits for air pollution sources. Most states already have an operating permit system. However, the new Federal operating permits will be issued under new rules which give the EPA veto power and allow all states bordering yours to comment. The permit system will be modeled after the NPDES system for wastewater. A significant

> **Air Pollution
> Federal Requirements (1990)**
>
> Title I - Nonattainment Areas
> - Potential new or tightened rules for various industry categories
> - Geographical area must show 15% reduction in six years — will get it from industry
>
> Title III - Air Toxics
> - Rules for industry categories and toxic type (e.g. general coating, automotive, boilers)
> - Rules for coating industries published in 1993 require Maximum Achievable Control Technology (MACT) = top 12%
>
> Title V - Operating Permits
> - Federally enforceable operating permits issued by the states (currently none)
> - Will be issued to all major sources and for air toxic sources
> - Permit fees will likely range from $25/ton up
>
> Title VI - Stratospheric Ozone Depleting Chemicals
> - Tax on 1,1,1-trichloroethane, CFCs and halon, 111-TCA phased out by 2002
>
> Title VII - Enforcement
> - New authority for administrative orders and field citations. Clearly defines person as senior corporate management not line operators
> - Bounty hunter and citizens suit provisions

Figure 16-10. Clean Air Act Amendments of 1990 are the foundation for issuance of new regulations in the coming years.

item in Title V is the collection of permit fees based upon emissions. The fees likely will range from $25 upward per ton.

Companies using degreasing solvents such as freon or 1,1,1-trichloroethane are already subject to the new tax on ozone-depleting chemicals. These taxes will increase from year to year until production is eventually phased out entirely. Title VI addresses ozone-depleting chemicals.

CAAA-regulated Organic Chemicals

Benzene	1,1,1-Trichloroethane
Carbon disulfide	Methyl ethyl ketone
Carbon tetrachloride	Methyl isobutyl ketone
Chlorobenzene	Methylene chloride
Chloroform	Nitrobenzene
Ethyl benzene	2-Nitropropane
Formaldehyde	Styrene
Hexane	Toluene
Methanol	Xylene

See regulations for complete list of air toxics and source category listings.

Figure 16-11. Rules pertaining to toxic air emissions that will impact paint operations will continue to be written in the coming decade. The regulations will affect not only actual surface coating materials, but such items as space heaters and cooling towers.

Title VII adds teeth behind the bark of the new CAAA. The enforcement abilities of the EPA and citizens' private suits are greatly expanded. The EPA will have a large staff dedicated to enforcement, particularly criminal investigations. More violations will now be considered criminal violations of the law, which place individuals at a risk for fines and imprisonment. Moreover, EPA will now be able to issue administrative fines after inspections without first providing a notice of violation (NOV) in initiating formal enforcement actions.

Citizens' groups will also be able to bring suits against companies for violations of the CAA when EPA or state agencies fail to initiate actions. Because of the "bounty hunter" monies available in settlements, citizens' groups and attorneys representing these groups use enforcement against companies as a money-making opportunity.

Facilities located in Canada will obviously be impacted by an entirely different set of environmental regulations. One noteworthy piece of legislation under consideration is the Clean Air Program (CAP). CAP, Figure 16-12, would restrict toxic air emissions by creating three control levels based upon the toxicity of the pollutant. In the proposed format, trace contaminants in paint solvents, such as benzene in aromatic blends, could be subject to severe control requirements. These Level I pollutants would be virtually eliminated by the use of the strictest control technology demonstrated anywhere in the world.

VOC equations

Many similar words or abbreviations are used in VOC terminology which can lead to confusion and even legal problems if noncompliance results. Always read the definitions in the applicable laws first. A list of some of the common terms appears in Figure 16-13.

Coatings contain volatiles and nonvolatiles. The volatiles include water, so the rules are often expressed as VOC less water. The volatile content is a percent by volume (gallon of volatile to gallon of coating) or a percent by weight (pound of volatile to pound of coating) or a weight per gallon. The VOC

**Air Pollution
Proposed Canadian Laws**

Clean Air Program (CAP)

Severely restricts air toxic emissions

- Level I - virtual elimination of emissions using strictest controls demonstrated anywhere in the world
- Level II - greatly reduced emissions using reasonable control measures and taking into account economic considerations
- Level III - reduce emissions where feasible and economically justified so that ambient levels are maintained below health effect levels

Figure 16-12. Even trace contaminants in solvents will be eliminated under laws proposed in Canada.

VOC Equations
Common Terms

Volume solids	%
Weight solids	% or lb
Volume volatile	% including H_2O
Weight volatile	% or lb incl H_2O
Formula VOC	Lb solvent only
Actual VOC	Lb solvent + cure
Transfer efficiency	%
As received	Unthinned
As applied	Ready to spray
Solids basis (RACT)	Constant volume solids
Capture efficiency	% VOC to controls
Destruction efficiency	% VOC in/out of control

Figure 16-13. Environmental regulations have imposed a plethora of new terms on the paint manager, who now must expand his or her vocabulary to grasp the rules of compliance.

content would exclude the water if present, but still includes "cure volatiles." Cure volatiles do not come from solvents, are not in the coating, but come from heavier, higher boiling resins which, upon heating, change their form, some evaporating. This is referred to as the actual VOC or test VOC. By contrast, the formula VOC includes only the theoretical amount of VOC resulting from the solvents.

Nonvolatiles or solids are also measured as a percent by weight or volume, or as pound per gallon of paint. The volume of solids in a paint is important in calculating RACT equivalency for controls, and in some regulations, it is an integral part of the calculation for compliance, Figures 16-14 and 16-15.

The VOC content can be expressed in terms of a wet gallon of paint or in terms of the emissions resulting from the application of a gallon of solids. Figure 16-16 depicts these two common equations. The first is pounds of VOC in a gallon of coating that has no water. If the coating has water, one must subtract that to determine compliance.

Many states and the EPA have started writing regulations which list the pounds of VOC resulting from the application of a gallon of paint solids, which

VOC Equations
Constant Solids

	High VOC	Complying
Usage gal	1000	?
VOC lb/gal	5.5	3.5
Lb/gal VOC	7.36	7.36
Vol % solids	25.3	?

Complying % solids
   $1 - (3.5 / 7.36) \times 100 = 52.4$

Lb VOC per day
   $1000 \text{ gal} \times 5.5 \text{ lb VOC/gal} = 5500$

Gallons solids per day
   $1000 \text{ gal} \times 25.3\% = 253$

Figure 16-14. Critical to calculating RACT equivalency for controls is the volume of solids for paint.

takes into account the efficiency of application. This transfer efficiency may be an assumed number stated in the regulation or it may be a number derived from a test of your equipment. A high transfer efficiency means you spray less paint to get the same amount of solids on a given part, and therefore have less VOC emissions. An added benefit in addition to buying less paint and emitting less VOC, is that there is also less oversprayed paint in the filters or waterwash.

Other health and safety considerations for the paint shop

Any discussion of regulations should include attention to some different hazards involved in the application of protective coatings. These make protective equipment and apparel necessary today much more than decades ago. Work in various painting operations in North America still finds a great deal of misunderstanding and violation of regulations concerning worker health and safety.

**VOC Equations
Constant Solids**

Equivalent complying gallons
253 gal solids / 52.4% solids = 483

Complying lbs VOC per day
483 gal × 3.5 lb VOC/gal = 1690

Equivalent control efficiency
$$\frac{(5500-1690) \text{ lb VOC}}{5500 \text{ lb VOC}} = 69\% \text{ minimum}$$

Figure 16-15. In some regulations volume of solids is an integral part of calculating for compliance.

**VOC Equations
VOC Content**

Lb VOC/gal - H_2O:

$$\frac{\text{Total volatile lb} - H_2O \text{ lb}}{\text{Gal of coating} - \text{Gal of } H_2O}$$

Lb VOC/gal solids applied:

$$\frac{\text{Total volatile lb} - H_2O \text{ lb}}{\text{Gal solids} \times \% \text{ TE}/100}$$

Figure 16-16. Two common equations for calculating VOC content.

I mentioned early in this book that nothing would ever be done to injure someone or to damage their health. No gallon of paint or pound of chemical has ever been worth that to me.

There are several potential hazards in the application of paints. Toxicity can occur due to inhalation, absorption, or ingestion of products into the body, primarily solvents. Today we are using coatings which may be UV curable. These will cause burns to the skin if not handled properly. Such products were not around several years ago, and even managers with several years' experience will have to learn new aspects of their trade.

There are maximum concentrations of solvents present that are allowed to be inhaled during an eight-hour work period without danger to an operator or supervisor. These are known as "Threshold Limit Values" (TLV) and spoken of in terms of parts-per-million by volume. Some have low toxicity, while others are very strong and have very low levels of TLV.

Unfortunately, most of the more common solvents used are aromatics; branch-chain ketones, trichlorethylene, and diacetone alcohol are all stronger solvents than the aliphatics and are photochemically reactive. Paints are limited in various solvents. All major paint companies are now actively formulating non-reactive paints that will meet regulatory requirements and still provide the quality levels demanded of the market.

There is no excuse for not using more environmentally friendly and safer coating materials today. To do so is a major failure of a finishing manager. Alternatives in powder, water, and solvent products are available as are more environmentally friendly cleaners and pretreatment materials.

We owe it to our operations to keep safety equipment ready at hand and in good working condition. Booth balance and other preventive maintenance is most important both for safety and good first-time-through quality performance. These little things separate the good from the poor performers.

It is my quiet satisfaction to know that many of the "nasties" were taken out of products in the shops I have been associated with long before the government mandated it.

It is obvious that little things such as ground wires, safety containers, eye wash stations, eye protection, uniforms, gloves, fire extinguishers, fire blankets, hoods and respirators, and other safety items are essential and must be maintained. We can all help in good housekeeping practices, removal of trash, and making certain that aisles are open and exit signs installed. Such measures as screening conveyors where they pass over any traveled aisle are simple and effective. Proper footwear can prevent injury from crushing. Conveyor screens and other hanging objects can be properly marked both on the floor as well as with streamers to alert people. A first-aid kit and some basic training can often be the difference between life and death or serious injury.

Develop a safety checklist. Do you have a plan for evacuation in times of a serious fire or natural disaster? Do you have check valves in paint lines to stop paint from flowing backward into tanks and overflowing in times of major power failures? Are there easily positioned switches in paint areas to shut down entire systems in case of major evacuations? Are you using proper explosion-proof or air-operated electrical equipment in your paint areas? Are there approved and operating weighted fusible link fire doors? Are there crash bars instead of knobs on other doors so people can exit more easily if smoke or other contaminants would hamper vision?

Many fires in paint areas have been caused by illegal or careless use of electrical coffee pots and other appliances. True, these are convenient and perhaps save a few cents for someone. Too often, though, someone forgets to pull a plug or the unit malfunctions, it boils dry, and there is a fire. I do not believe such appliances should be permitted by any responsible manager (nor, in many areas, do local fire officials). It is not a case of being a "hard" manager. It is merely common sense.

There are those also who believe paint ovens are there for heating food. Some very toxic products can be generated in an oven during curing. These can wind up in your food, with obvious consequences. Moreover, not only are humans at risk, there have been cases where containers have exploded and products were rejected as a result.

To avoid this, a specific area for eating should be provided with provisions for heating food properly. Refrigerators for food storage should also be made available. Drinking fountains, wash-up areas, and clean lockers and restroom areas, of course, are a given, since they are required by law.

Health and safety considerations must be a large part of any well managed operation. The six paragraphs listing the numerous items that can be monitored and performed by every person in the paint operation are common sense things that do not require specialized training. An informed team of people who place the appropriate level of importance on these areas will go a long way toward improving conditions and performance.

The real professional can make just one quick inspection of a paint area and tell if management cares and is stressing the fine points of working environment as well as production. Experience has shown that 99% of the time, emphasis on these items of health and safety will also find emphasis on producing quality work with competitive costs.

CHAPTER 17

Supplier Relations/ Responsibilities

Examination of resources from both outside and within the organization can offer better utilization and performance if we move away from the former combative relationships with suppliers.

One of the more significant changes that has taken place in the past few years is in buyer/supplier relationships. Conventional former practice was to have three potential suppliers of everything needed in a paint shop. This was true for direct as well as indirect materials. About the only exceptions were certain service parts for equipment that fit only one product. Even then, it was common to have more than one potential source for those parts.

It was quite normal for purchasing policies and accounting practices to demand competitive bidding. This took an inordinate amount of time to meet with all prospective suppliers and review their products. Then, more time was spent testing and evaluating each new product purported to be better than the ones in use. All this was needlessly redundant, but the bureaucracy of business placed a great deal of power into a few hands that controlled the purchasing of materials and services.

Unfortunately, this often led to abuses. Suppliers wooed buyers with entertainment and perks to win orders and contracts. This practice led to acquisition of materials and services of questionable quality, which invariably manifested itself in the shop's performance.

In those days, the paint industry was populated by a disproportionate number of people of questionable character who thought nothing of taking kickbacks or exercising preferential buying. This is not to say that everyone fit this description, but the industry in general did not have a good reputation 40 years ago when I first got into this business.

Fewer suppliers, longer relationships

We tend today to prefer more long-term relationships when selecting suppliers. Products are not changed as often or as easily as they were in the past. We now look at the total cost of doing business and what role products and services play in achieving our corporate goals. Decisions are made on a much more professional level today. Evaluation and trials of suppliers' products still take place, but there is more initial screening, and even off-site testing in application centers is now provided by suppliers. By the time products are put through this testing, most of the unknowns are known and there are fewer surprises when actual line tests are performed. Also, astute managers today avoid using their everyday production lines as laboratories. By using suppliers' facilities for testing, they avoid production interruptions and preclude prejudicing the opinions of their workers. Proving the product outside of the shop ensures that you will bring winners into your operation. This, in turn, creates a winning, successful atmosphere and builds confidence within the organization.

If this type of evaluation is not feasible in your company, ask the supplier(s) to provide written certification of quality and performance for each batch of products received. This is often the most viable alternative and is being selected more and more by large operations.

All in the family

With the advent of enterprise integration, suppliers are now part of the product team, brought into the inner circle of paint operations when chosen as the source of supply. Data and information are shared at levels never considered just a few years back. As members of the product team, suppliers are made part of long-range planning and share in the responsibility of improving quality and cost in a continuous manner.

This level of involvement permits suppliers to contract for raw materials over a longer period of time, control their inventories, participate in JIT ventures, train and provide better support personnel, and enjoy greater stability in their internal operations. And through electronic data interchange, instant paperless communication can be provided.

Another new and beneficial practice is single sourcing at both plantwide and corporate levels. Greater use of common equipment and materials in different plants within large companies has supplanted islands of technology and automation. In the past, it wasn't uncommon to find the same type of product finished a dozen different ways, even though common sense said that some of that product was not going to have the same quality, and different costs would be associated with each method.

Paring down suppliers will mean a stronger supplier team if the team is selected carefully. Naturally, costs will come down owing to the elimination of redundant activities. Responses to service support requests will be quicker and better, and, because the account is worth more, the supplier has more incentive to "go the extra mile" to not lose the account. The other side of this managerial coin is the commitment of all parties involved to learning the roles of the others.

For example, there are three principal parties involved in ensuring that coating work is produced properly: the material supplier, the equipment supplier, and the user of the products of both. It is the role of the material supplier to deliver to the user a product of uniform quality based on the specifications of the user. In addition to this, the supplier must be prepared to offer and provide services in the various chemistries needed to supply the user proper solvents, viscosity, and speed of movement through circulating lines. Suppliers will also assist in inventory turnovers and scheduling, return or disposal of empty containers, and in educating the manufacturer in technology trends. The product should be certified as correct to avoid performing redundant in-house receiving inspection testing. Suppliers should participate in training personnel and be accountable for participation and correction in the event field failure occurs. Standard operating procedures and MSDA sheets should be provided.

The responsibility of the equipment suppliers is to provide the best possible equipment to perform the job, to instruct the user in the proper use and maintenance of this equipment, and to advise the user of advances in the application or finishing technology trends. The user should always be provided parts lists and maintenance schedules in binder form that permits pages to be easily replaced to keep such information up to date. Both parts and/or service people should be easily and quickly available in the event of emergencies. Good preventive maintenance and control will avoid serious forms of breakdowns and limit such emergencies, but nonetheless, provision for such contingencies should be made.

The responsibility of the user is to use, maintain, control, and supervise the materials, equipment, and services as prescribed by the suppliers in a professional, continuous, and ethical manner.

The selection, development of personnel, control, and management must come from the user's own efforts. Material and equipment partners cannot be held responsible for the actual performance of personnel, the setting of operational conditions, and continual maintenance of facilities. This is the responsibility of paint shop management.

Outsourcing management

There have been efforts in recent years in some industries to sell and provide the service of management for a paint operation directly from the supplier of paints. They will take over the actual operation of a paint area and manage it, and, in some cases, companies will contract the maintenance service needs.

There are cases where this has worked and others where it has been less than satisfactory for any number of reasons. My personal opinion is that the paint manager keep an open mind and weigh each case for its merits or negatives. I do tend to come down a bit more on the side of not letting an outsider take over the management of a paint operation.

Management's ultimate role

Management's first job is to manage its activities and train people to perform. It must give the ultimate direction and goals under which it is to function. While we should carefully consider making our suppliers a part of the planning and operational team, it is still the responsibility of the company management to make the final decisions and then provide the means to perform.

The ultimate test is performance. Achievement rather than knowledge remains both the proof and goal of any management unit. Peter Drucker tells us, "Business is a human organization, made or broken by the quality of its people. Knowledge is a specifically human resource. It is not found in books. Books contain information; whereas knowledge is the ability to apply information to specific work and performance."

Management must make a productive enterprise out of its human and material resources. The examination of these resources from both outside and within the organization can offer better utilization and performance if we move away from the older combative relationships with suppliers. A highly selective and more inclusive approach has shown many positive benefits can be achieved as long as each of the three parties involved understands and carries out its respective responsibilities in a coordinated fashion.

Even though a much higher level of inclusion is important under new approaches to supplier relationships, it is also good practice to periodically review your needs to determine if they are still being met or have changed. Technology seldom locates at the same sources each time over long periods of time. A sensible review of technology direction throughout the potential supplier base every three years is both logical and necessary. This does not necessarily mean you start over. It means you look at your present supplier, who had previously earned your business, and give that source first opportunity to continue to meet your needs before you look elsewhere.

Vigilance and fairness

Even though supplier relationships are on a higher professional plane today, there can still be abuses. One must be on the watch for indicators that something improper may be happening: Is someone living a lifestyle beyond his or her means? Are certain test data or panels lost from time to time? Are there delays in getting approvals for materials evaluations?

There should always be a stated, or at least implied, code of ethics. There should be a means to convey to executive management allegations and proof of violations without fear of recrimination or breach of confidentiality.

On a more day-to-day basis, there should always be someone to serve as a "court of appeals" for suppliers. There will be times when some matter needs to be heard and evaluated and an answer given. This court of appeals may be either a single person who has a solid reputation for being fair and impartial, or a small group of people. My experience is the smaller number involved, the better, for a variety of reasons. This keeps the matter more confidential in most cases and leaves no doubt who made the decision.

Good suppliers just want to know that the playing field is level. They ask only for the opportunity to participate. They want and deserve to know the rules and how the buying decision is made. There will be disappointments. Not everyone can be selected. The good suppliers will keep coming back if they feel they were given fair and professional treatment.

Any attempts by either party to operate outside these logical parameters is suspect. My experience has shown that any supplier relationship outside the parameters of professional conduct is built on quicksand and not in the best interest of the company, the product, or any part of the paint operation.

CHAPTER 18

Sabotage and Other Negative Communications

In any shop situation, there exists a small group of people who are a potential source of sabotage and other counterproductive activities.

For want of a better world

Without question, virtually everyone would like to work in a world where there are good communications, continual harmony, and a trouble-free existence with stable employment.

Unfortunately, Utopia does not exist. As long as people have individual personalities, they will have imagined and real grievances that will affect performance. Different values, different wills, and different attitudes create differences of opinion on the shop floor. I debated whether to include a discussion on human behavior in a book about paint shop management. Would it present a negative statement to the extent that such a sharing would defeat the other positive suggestions in the book? In the end, I decided it was important to include these anecdotes because similar events still occur today. Perhaps this discussion will help prevent such happenings in the future. Perhaps it will impress on management that they can be victimized quite easily by someone with strong enough intent to do so. Whatever the result, to be forewarned is to be forearmed. The following collection of potential problems represents just a few of the incidents of negative communication I have observed in my career. Some were intentional, some were accidental, some were the result of negligence or carelessness.

- Styrofoam cups discarded into paint circulating tanks. These dissolve into small balls, clog up valves, and spray as dirty paint even with good filters.
- Silicone compounds applied to surfaces or added to coatings that create fisheye cratering, Figure 18-1.
- Incompatible lubricants introduced into bearings or agitation equipment, resulting in cratering and paint adhesion loss.
- Dry powdered soap from restrooms introduced into gloves of painters and shaken out onto product surface at various times and locations during a shift.
- Kinks in air hoses supplying pressure pots. When pots are placed on the hose after leaving the fluid valve on, no paint will come from the pot because the air is choked off. However, if the pot is moved, the restored air will cause paint to come from the open valve and spray everything in the area until it can be stopped. Never approach a leaning pressure pot without surveying it for an open valve.

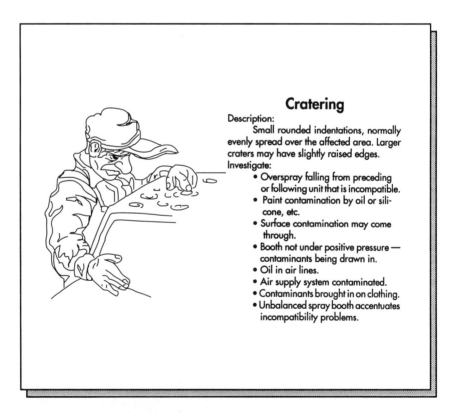

Figure 18-1. A perennial problem of painters, cratering requires much unnecessary rework.

- A small, thin block of wood placed under electrical control switch buttons so conveyors and other equipment cannot be started.
- Equipment and materials hidden in trash containers. These are excellent places for people to hide paint guns, sanders, and supplies so an associate on another shift can steal them when the line is not working.
- Idlers in hideaway places. People use these to sleep in, as well as store and use alcohol or drugs. Take note of unusual location of wires or antennas. Notice if storage containers are never moved. These are all signs someone has made them a "nest" in which to hide.
- Cigarette butts or other debris in stair wells, etc. in off-shift areas. These are probably sites where workers congregate for any number of reasons: gambling, drinking, etc. There have been instances where workers have lost large portions of their paychecks at the plant before even going home.
- Introduction of common non-paint products into painting systems, such as e-coat lines, in retaliation for automating a painting operation. It is easy to find items such as antifreeze or fuels around many plants that can be easily introduced into systems during weekends.
- Introduction of lacquer paints in small quantities into enamel paint circulating systems. This, of course, will cause great kickouts of resins and major down time.
- Bulky clothing. Look for changes in people's appearance or clothing, especially in winter. A good deal of theft can occur by workers wrapping their bodies with materials or carrying them in pouches under clothes.
- Filled thermos containers at end of day sometimes can mean theft of expensive isocyanate hardeners or even paints stolen in small quantities over a period of time.
- Nighttime theft. Keep areas around plants lighted. Paints are sometimes lowered from the roof of a plant to friends waiting outside for the materials. Look for someone involved in an outside business activity.
- Thin layers of water on the contents of full drums of paint. The water will likely be blamed on a loose bung and the entire drum of paint declared as scrap. Paint is then picked up or sold to someone for little or nothing. If handled carefully, the drum can be taken to another site, the layer of water and some paint pumped off, and a majority of paint recovered.
- Guards and other security personnel moving property from the site or breaking into cigarette and concession machines or coin changers. These are the greedy forms of theft.

- Vending machine theft. Look for unusual holes in vending machines on their back sides. Do not overlook the use of templates made to drill holes to activate change makers or obtain products by triggering through these holes. These are the patient, non-greedy thieves.
- Nighttime illegal entry. Use selective key-type locks that have special key blanks. Many auto keys will unlock other commercial forms of locks. (I once had an ex-convict working for me who had been trained as a locksmith while in rehabilitation. He showed me how easy it would be to carry the place away by someone who knew the trade. Fortunately he was going straight and continued to do so.)
- Preparations for theft. Plants that have no roofs on storage cribs, etc. are subject to more theft and sabotage than those with them. Look for placement of tables or cabinets along their walls that offer a convenient opportunity to scale the walls. The same goes for the inside.
- Stealing of cores and other reworkable castings such as engine blocks, etc. or old paint equipment is a common occurrence. Often, these are damaged intentionally so they are available either by theft or buying at scrap value and then remanufactured.
- Theft of consumables. Check the budget to determine the average cost per employee for supplies such as soap, toilet paper, coveralls, shop towels, etc. Compare this usage to your personal experience at home. These things can either be taken from the work place itself or never delivered to the plant at all.

People are capable of some strange and resourceful antics and are motivated by an array of reasons. These actions have come from both management people and floor workers. Each is capable of being motivated in a negative fashion.

Overtime fraud

Probably one of the most insidious forms of theft comes in the form of fabrication of overtime in plants. When this is permitted to happen, control of performance and costs rapidly deteriorates. Once people get used to making this extra income, they find it difficult to return to normal. People who otherwise would not take a penny from anyone will play the game of overtime fraud and not feel they are stealing or sabotaging anything. I have seen people double their wages by this method. They did not have to perform any better. They just had to show up. It is not uncommon to find first-level employees making more money than high-level personnel with much higher skills. Such an operation sees just about all incentives to perform drop out of sight.

In this scenario people become tired before long and really end up not putting out any more product in the additional hours compared to normal. They work slower because the body won't perform at a very high level for long periods of time.

Eventually something will occur when still more is required, and then companies are almost held hostage or blackmailed into giving even more considerations just to keep people working the overtime. For those of us who have been around industry for a long time, this situation is quite recognizable. It is the slow creeping poison of destruction for many a company. People know it is wrong, yet they are still willing to take the money. It may work for a while, but eventually they are going to have to account for it. It is like a person falling off a 100-story building. About halfway down, he or she looks around and says, "So far everything is okay," but a rude awakening is in store.

Management's Tower of Babel

Many times negative behavior has its genesis in an urge to get management's attention and begin some form of negotiation to satisfy a complaint. Formerly, management came from the ranks and understood this type of message. Managers would try to find the problem area and react to solve it. It was the language used by labor and management to communicate, and everyone understood it.

Today, many management people come directly from college and do not have the benefit of this experience and communicative style. This poses a real problem. New managers are purposely targeted and "tested" to see how they will react in a situation. If they don't react properly to stop an activity needing correction, that activity will normally continue until it becomes an accepted way of doing business.

Clearly, the inexperienced management person does not favor this scenario. This wasn't taught in school. The normal reaction is to look upon such actions as a very negative thing and simply plan to find a way to eliminate the "problem" by outsourcing it. But this leads to loss of jobs and still further breaking down of relations between management and workers. This change in management makeup and the resultant change in approach to problem solving has been a hard lesson in communications for labor leaders, also. They have found the old style doesn't lead to resolution, but to loss of jobs when they target an area. These labor leaders do not realize that their former successes were now simply "drawing a bulls-eye" on a situation. By pointing out a problem area to the contemporary management, labor leadership was in fact fostering job losses rather than communicative negotiations.

There is still another twist on negative communications. Often, when management is promoted from the ranks, the new managers are former labor leaders: they had shown leadership and management qualities in those former functions. Since these former labor leaders had created and negotiated most of the work rules being used, they believed in them, having been a part of their creation.

The potential question is, do they believe in them so much that they will not change the operating procedures, even long after technology or something else has obsoleted them? It is possible that they will find every reason to continue the old ways and not be flexible to new needs.

Business today has been reducing both the work force and the management personnel to operate in a leaner mode. They do not wish to add full-time personnel because of an unknown business climate, perhaps, or new governmental regulations and taxes, or cost of the scheduled benefits associated with full-time employees. Instead, current staff is on scheduled planned overtime. It is my fear that, although we will see some short-term benefits and profitability from this, it will become the insidious killer of productivity in the long run. The present-day employee is making more but enjoying it less, and the new entrant into the job market cannot find employment at any meaningful industrial job.

While we do not want to believe anyone would have any negative thoughts about our operations, truth and experience have shown that such beliefs are naive. There are always those who are a potential source of direct or indirect acts of sabotage and negative communications. How management handles this small percentage will dictate how most of the others in the organization react.

CHAPTER 19

Winning the Professional Way

Effective managers invest in people and help them grow professionally. Then they reward them for that growth.

In the preface to this book, I stated that one of the greatest problems in my nearly 40-year career was that of acquiring qualified management for paint operations. It was both my hope and intention that each chapter would provide at least one thing which could be helpful in becoming a better paint manager.

There are hundreds of ways to fail in the business of finishing. It is important that you not learn them all by experience. That would be a long and harsh road to travel. In fact, it is doubtful you would be around at the end of the journey; either you would have been dismissed as a failure along the way, or your health would not have stood up under all the pressures and disappointment.

Substance, style, or gift?

What we have to do as good managers in the business world is to learn from the successes and failures of others. Couple this with your successes in limiting personal failures, and you will grow every year of your life in your management capability.

I have found no exact science of management. It is a combination of substance and style. This combination will vary and change over the years because we live and work in an ever-changing environment. The astute person recognizes and applies proven basic principles to these changes. He or she deals with change as a fact of doing business and learns to manage and control it.

Remember, a good cook knows how to vary the "M&Ms" which make the ingredients of a business operation. There are indeed several roads that lead to the town square, yet surprisingly, it is not so important to have any specific road to the square. It is often left entirely up to the individual. The goal is just to get there. In other cases there may be much more specific guidelines as to just how, when, and where you are expected to arrive. My own experience involved much more of the former and fewer of the latter. Those giving the goals expected me as a responsible manager to choose the appropriate way. They empowered me and shared the authority to do it. And these, I got the old-fashioned way, by earning it, beginning with little things and working up to larger things.

It is necessary to learn the difference between leadership and management. Sooner or later, a person will have to do some of both. Knowing when to provide which comes with experience. It comes more naturally to some than others, but a great deal can be learned by a person wishing to excel. We learn the lesson of desire early. Be sure the fire of desire warms and nourishes you and does not engulf and destroy everything.

The people connection

Management is getting things done through and with people. Never forget that the key is people. People are the difference, and smart managers surround themselves with good and true people. But they don't stop there. Good managers continue all their lives to help people grow in capability and reward that growth.

Management's first function is to run the business and put economic performance first. This is the way it justifies its existence. There are, however, a number of social contributions that help people, the community, and society in general if management produces economic results. None of this will be furthered if the manager fails as an economic performer. In my mind, a company has an obligation to contribute beyond the economic arena. We need to participate and bring our talents to the community in order to have a balanced and satisfaction-filled life. The seed corn philosophy works not only in the business arena, it works outside as well. People must put something back into the world. We cannot always be takers.

Only people are capable of enlargement. They cause transformation to take place. Only people can grow and develop. Everything else is subject to laws of mechanics. All other resources can never have outputs greater than the sum of their inputs.

Managers must learn to manage both up and down the organization to be most successful. We must learn to manage managers because this is what turns a collection of resources into a viable enterprise.

A final function of management is to manage work. This means making work suitable for human beings and enabling them to work in a productive, safe, and effective manner. Every thought and function should be considered from both the present and long-term survival of the organization. Without this, there is no balance and harmony. Resources will be endangered, damaged, or destroyed.

There must be desire and dedication. A price is attached to becoming a high performer at anything, and this price is paid in becoming a good leader and manager. The greatest lesson I have learned in my career is the more one gains responsibility, the more one becomes a servant to all they direct. You must satisfy the wants and needs of all those in your organization. Leadership and management involve serving others.

To be effective managers, we must know who we are and what we are trying to contribute. This is done through a series of philosophies which enable a strategy to be developed. Painting a product is such a philosophy. It is like an iceberg; only a part of the whole is readily visible to most people. Most of the required substance to perform is below the surface and not visible to the layman. The paint and finishing discipline contributes in many ways to the total manufacturing operation. It begins when a product is being created, carries all the way through the many steps in manufacturing, and continues out into the eventual reliability and satisfaction provided the customer. A failure to recognize the total contribution paint can make is a gross failure of any manager.

Ethics in management

A leader or manager has a special responsibility to maintain a high ethical standard. There should never be any case where personal position is used to gain a lifestyle or other tangible benefits at the expense of the organization that person represents. People want to know that their futures are in the hands of ethical and responsible management. In an age of leap-frog technology advances, this is especially true. Workers want to be assured that technology is not being employed against them and their need to have gainful employment.

Once leadership has established the direction and tone of an enterprise, management must make the assets under its direction grow. There is a fiduciary responsibility to protect assets and enlarge them.

Good managers teach their people how to fish. This will make them capable of supporting any need required in either their personal or professional lives. It is a lifelong and never-ending process of learning and then teaching to others. Remember, people are also keeping tabs on you, so teaching them well is important. Your slightest deviation will be noted and could cast doubt on your credibility.

Becoming lean

Industry today is thinning out middle management to reduce costs. This makes it necessary to have line people begin to take on some of the decision-making roles formerly performed by middle management. It is necessary to share much more information than in any time in history. We must begin to empower workers, train them in their new and expanded roles, and then trust them to perform. A modern manager is flexible and readily adjusts to change.

To be effective, managers today must be much more knowledgeable about other disciplines in the total organization...marketing, engineering, accounting, purchasing, and product reliability. They must know about health, safety, and regulatory requirements. In addition, familiarity with organizational development is necessary to maintain order and operational control. Managers must be willing to change the organization as technology and other variables obsolete current techniques.

An astute understanding of people is essential. People have distinct characteristics that influence their reactions in the workplace. Understanding these differences will greatly assist managers in selecting, training, and building the most effective organization.

There must be a dramatic change away from the accepted concept where people feel entitled to everything. Workers must have the opportunity to dream and work toward those dreams, a part of which is done in the workplace. Like all benefits, this must be earned and not assumed. Benefits or other rewards provided without increasing performance will only foster a slow death for the organization. Providing rewards without demanding performance puts the entire enterprise at risk. It has happened in the past and is in part responsible for the downward spiral of American industry. But it can be reversed with leadership and management that inspires people to perform in lean work environments. American workers are not the lazy and unproductive lot portrayed to the world. They simply have been deprived of flexible management willing to change with the times, share decision making, and invest in technology.

Performing in business as well as in our personal lives is a matter of putting a lot of little actions together. These little building blocks become the foundation of everything we do. Success comes from doing the right things in the right amount and at the right time.

Do not become overwhelmed with complexity. Learn to look at any plan or action as a series of small parts or steps. This will permit more flexibility and present more alternatives, and you will have a greater depth of understanding for the whole when completed. Follow the lesson of the turtle: slow and steady in nearly all instances will put you by the rabbits.

Strategy and planning

There is only one reason to have a paint operation. It must help meet the company's strategic plan and goals. To do this, a paint manager must participate in the business planning process and be knowledgeable of the basics of a business plan and the way funds are handled. Knowing the accounting system and how it impacts on the paint area is imperative.

Once a business plan has been established, the manager must create a sound budget. This should be done in such a manner that all people involved will buy into and perform in a manner to guarantee its integrity. Good managers then monitor all the variables possible to control and improve costs. Results of budgets and efforts to control performance should be reviewed and shared with all personnel on some type of regular basis. A pledge to operate under control is not to be taken lightly. It is another important opportunity to create credibility.

Winning managers learn and use the new methods of ISO 9000 and SPC as part of their operations to improve quality and performance in a continuous manner. Such methods are becoming increasingly necessary to doing business today in most areas of the world. The trends toward such certification and control of quality are growing rapidly.

Such methods have become the backbone of troubleshooting and problem solving. They work for both identifying and resolving problems. They can also be used to prevent problems. The old saying about an ounce of prevention is worth a pound of cure is certainly in evidence in this modern-day use of statistical control.

Maintaining a profitable operation

A good manager will support a strong maintenance effort. A sound set of operating procedures ensures smooth starting and shutting down of systems. A strong preventive maintenance policy can be compiled from data collected along with supplier and maintenance contractor support.

Maintenance today extends far beyond the typical areas found in the past. It is a major contributor toward employee health and safety, quality, facility protection, and environmental regulations. There is not a gallon of paint or pound of chemical worth getting someone hurt or sick.

As managers, we are charged with selecting suppliers and maintaining relationships with them. Modern approaches more and more involve them as partners in an expanded role in helping to operate a paint area. When properly applied, added technical knowledge and training can be a great deal of help. Such relationships carry a lot of ethical responsibility: they should never be used to gain an enhanced lifestyle through gifts and excessive entertainment. All good companies have policies guiding the ethics and code of conduct in this area. These must be adhered to and enforced.

Some of the most successful people I know are contrarians. This means they have a way of doing things that appears contrary to what the majority of people might do. It doesn't mean they are crooks when everyone else is honest. It simply means they tend to spend money — albeit prudently — when no one else is doing so and save money when others are spending it. When business is good, they work very hard to produce product and serve customers' needs. During good times, they try to get all the cash possible into the bank. They run a tight ship during this period and avoid spending sprees. They do not disrupt their operations with major capital programs. Employees commit to work extra hours and few additional personnel are added.

When business is lean, they work hard at producing product and serving customers' needs with whatever business there is. They do not suddenly put a freeze on spending as many normal businesses would do. These managers use this down period to renew and get ready for the next up cycle. They do not lay off their trained and loyal personnel. Instead, these managers put their people to work cleaning or rearranging the plant. This is the period when capital monies are spent, not reduced. Such a down cycle in the business world presents a better opportunity to obtain more competitive bids on goods and services. There will be shorter waiting times for contractors to make facility changes because those companies, too, are trying to keep key people employed.

Rearrangement during a down period presents less risk to the organization than attempting to do so during an up cycle. With the well run organization, there is not a problem with money because cash was built up during the up period. If additional money is required during this effort, borrowing is a good option because interest rates tend to fall during depressed times.

When the refurbishment is complete, the manager has a facility capable of responding to higher demands and quality more efficiently. Workers display more loyalty to the company because they were not sent to the unemployment line like their neighbors. The company has helped the general condition of employment in the area by not laying off its own people, and it provided additional work for contractors and suppliers. It is a win-win situation. It is not a traditional way of thinking, but it works. I have found this concept to work in both my personal and professional life. It is basic, doable, and much more benevolent than conventional approaches.

The art of finishing

The finishing business has been a wonderful and fascinating career for me. I have strived to make it a more professional and performance-contributing part of the overall manufacturing mix, and I take a certain amount of justifiable pride when I see the modern paint lines today compared to what they were four decades ago.

It is my hope that this book will have increased your awareness in many areas...that you will be challenged to add to the improvement and professionalism in the paint business of the future. One person can still mean the difference between success and failure.

No one can take knowledge away from you. Seek it out. Knowledge, combined with desire and motivation, is the fuel that drives accomplishment. Anything done to excess will fall short. Learn to balance your activities and keep a perspective in everything.

Do your very best to provide an environment where people are not manipulated but given positive reinforcement for a job well done. Reward completes the performance circle. Lead people, manage things.

If we continue to do the same things, we will keep getting the same results, but the same results will not suffice in the global business environment of the twenty-first century. This means a smart manager has to adapt constantly. Continuous vigilance and continuous improvement are the strategies of competition today.

Signing off

To the young managers in the finishing industry, we do not know where technology will take us in the future. Space travel tomorrow may be as common as air travel is today. If that occurs, do something for me when you are out there high above the earth.

Though someday my time here will come to an end, I don't expect to leave the finishing business. Peer down toward earth, if you will, and look at the rainbows. I've already applied for my future job of helping to paint rainbows. Look carefully and wave as you go by. It's a beautiful finish, isn't it, and it is compliant.

Thank you, and good painting!

Et Cetera

One of the most effective ways of communicating is through humor and anecdotes that paint word pictures to people. It works at all levels. There is a time and place for this type of communication, and a manager's selective use of it may often play an important role in bringing opposites together.

There are also times when propriety argues against using humor. That is what is so fascinating about communications and various styles of management: there are many ways to reach the same goal. The art is to recognize and perfect more than one; the science is to discern which to use at the proper time and place.

Part of being credible is the ability to impart to others the capability to "read" and know you, so pick the anecdotes that reflect you as best you can. When you are all things to all people and never reflect some fairly high level of consistency to others, workers will seldom welcome you into their circle. This prevents a manager from expanding his or her level of influence on things that have a great effect on them.

After 40 years, I have collected a number of experiences in the industry. It is my hope that by sharing some of them, you will gain some insight into how humor can be used to communicate. The time taken to tell a story often cools and defuses situations and restores calm, and blessed are the peacemakers.

Bird dogs and basic lessons

Working with bird dogs and training them has been nearly a life-long hobby of mine. There is a classic story told more than 60 years ago by a gentleman named Horace Lytle, who used to be the gun dog editor for *Field and Stream* magazine. His story has been passed down through generations now and points out what can happen when communication breaks down.

It seems that it is accepted good practice to never make a pet out of a professional hunting dog. The theory is you should only take them out of their pen when the dog is going to be either trained or used for running down birds. In other words, keep the dog focused on what it is supposed to be, a hunting dog.

One very cold winter, a champion female bird dog had a litter of pups, and one of them was not doing well. This runt of the litter was going to die, and normal practice is to let that happen so as not to taint the breed. This one time, however, the owner broke his own rule and took the little pup to the house to try and save it with personal attention.

The man and his wife made a little box behind the kitchen stove where it was warm, and personally hand fed the dog on a regular basis from a nipple and bottle till the little runt improved.

Since it was in the house, it soon became obvious the puppy would have to be housebroken, so the owner set about attempting to train him. He tried everything. He took the pup out on a regular basis, tried newspapers on the floor, even rubbed the dog's nose in the subject of his training. Nothing worked.

By this time, it was beginning to be spring. The windows were now open with the warm and fragrant air coming into the house. One evening, the owner was sitting in the front parlor reading the newspaper. The now much larger pup came into the room and promptly demonstrated on the carpet that the training had been in vain. Well, that just about broke the patience of the owner. He jumped from his chair, grabbed the dog, rubbed his nose in the mess, and threw him out of the window.

The dog finally got the message. Or at least *a* message. Now the pup comes into the front parlor, does his business, sticks his nose into the mess, and jumps out of the window!

It has struck me that we often communicate the wrong desired result when we deal with others. Our intentions are good — like in the case of saving the little pup's life — and we work diligently and long, but with little success. Then we lose control, and that is what people pick up on. Unfortunately, it is the wrong result. We still have the mess on our floor.

Much ado about nothing

Sometimes we confuse activity with accomplishment. I have done some singing and entertaining as a hobby during my years here on earth. Have even been paid for it at times. But entertaining doesn't mean you leave everyone happy and satisfied.

There is a story about someone entertaining for a rest home where sick and elderly people lived. An old and infirm gentleman was one of those in attendance. After the show, the manager asked the singer if he would take a moment to see the old gentleman and speak to him personally. The singer stopped and said he hoped he got better. The old gentleman replied, "Son, I hope you do, too."

This can be an illustration of how we often compare apples with oranges in terms of considering performance. How good is good or compared to what can be different things to different people and managers, based on their experiences and expectations.

Color by the numbers

There was another time in my life when a failure to communicate produced a result different from what I expected.

Early in my career while working at a front-line management level, my company received a great many orders calling for special-color paints. It was winter, and when the paints arrived from the supplier, they were stored outside. This was a very poor situation. We could not locate the paints, and they were cold and covered with snow when we did locate the right skid of buckets.

My supervisor did not want to spend any money for overtime, but finally we had so much downtime and so many quality problems, that he relented to my suggestion to clear out an area in an indoor warehouse where the paint could be arranged, stored, and protected in a safe manner. He gave me one man to do this in one eight-hour Saturday.

This plant was strongly unionized and had rules about who got to work the premium time on Saturdays. It turned out that the person eligible was truly unique. He was good natured, hard working, but just a bit slow on the uptake.

Knowing this, I took care to take him under wing and thoroughly instruct him on just exactly what had to be done. He was to bring all the pallets of buckets into the empty warehouse, deband the buckets, and localize their positions based upon families of colors. Paint in our company was coded by color families. He then was to further sort them into individual colors by number into separate rows starting at the lowest number and leading to the highest number. All blacks began with a zero, tans with a one, reds with a two, orange with a three, and so forth. The next numbers were the respective color. Fairly simple.

The gentlemen told me he would get it done. In fact, he told me to just go ahead and do something else as he didn't really like to have the "boss" or anyone look down his neck while he worked. I left to do some other things feeling quite proud of my instructions and confident the worker understood.

About two hours later I decided to go check things out. Upon arriving in the paint warehouse, there was this man sitting high atop a huge cube of stacked paint buckets and calmly taking a smoke break. He greeted me, said he was all done, and asked what he should do next. I was dumbfounded because this was not what he had been told to do at all. I asked, "What are you doing smoking in a paint warehouse?" Whereupon he said, "Oh, I forgot," and promptly hopped down off this huge block of buckets and put out the cigarette.

Next I asked why he had not arranged the paint in rows by family and specific number of colors. His reply was, "It's the same color," and I knew better and told him so. He insisted, "No, it is all the same color," and promptly showed me the number on the can, which read 2069.

I didn't know whether to get angry or laugh. What he had pointed out to me was the number 2069, all right. Unfortunately, he hadn't looked far enough on the bucket stencil. You see, the address of the plant was 2069 Lagonda Avenue. He thought this was the paint color. Of course, it was on every bucket received. As long as I live, I will never forget the address of that plant or forget to stay a bit longer and be sure I have communicated properly.

A little more persistence on my part in instructing him, and a bit more initiative on his part in double-checking numbers could have avoided this communication debacle.

The following story is one that will explain the difference between the amateur and the professional in so many cases. It has a great many examples of qualities that successful people have and how it can serve to train and motivate people around us. I call it the following:

It's what you do when you don't want to do it that counts

My family and friends often ask me if I ever get tired of all the problems. I reply, "Usually not, but if it happens, I just keep on trying. It's a compliment when people want you to help with problems."

When people wait for the right moment, they never become successful at anything. What most people never learn is that if you do something often enough, you are able to do it even when you don't feel like it.

We always talk about bad habits in our daily business. What we need to do also is to talk about good habits. Good habits work for us as strongly as bad habits work against us.

A ball player, a musician, or an actor does not always feel like performing. But being professionals, they do so anyway. The chief difference between the amateur and the professional in any field is not necessarily talent and skills. It is attitude, based on training and habit.

When I was younger, I competed in a number of amateur and semi-professional sports leagues. I often found players who could throw the ball as fast as any professional. They just couldn't do it at a high level over a prolonged period of time. They weren't dedicated to practice.

Amateurs only play when they feel like it, and their performance can range from super to lousy. This impulse reaction is why their performance levels are so erratic from day to day. The professional will win most of the time simply because of practice and consistency. He or she plays nearly the same time after time, and even rises to greater heights during important contests.

This pattern is the result of training and habit. The same habit forming can kill performance. Things done instinctively through habit are always easier than those done consciously. It is the same phenomenon that slows the typist who looks at the keys.

We have to train our workers and ourselves to do the right things every time so they will become instinctive. This applies to feelings as well as body reactions. Unless feelings are trained, the body won't react in a consistently professional manner.

Conveying the truth

At one time in my consulting career, my services were called in when a technical matter concerning the line speed of a certain paint conveyor was in question during the second shift. The union charged that the company was speeding up the line in violation of the contract. Management tried in vain to settle the dispute, and both sides had agreed to let me try to settle the matter.

It seemed the problem was one of the conveyors running faster at one end than at the other. Now there is no want of stories such as this in industry, and they are generally looked upon for what they really are-someone's joke. It really could not be happening. But it was, and the steward was really serious.

The conveyor was long, winding its way all around the plant. It went up and down several inclines and through load and unload situations. At various times, parts of different sizes and weights were hung on the conveyor or removed from it.

Some "slack" showed in the conveyor drive and takeup, this to prevent shearing pins where stresses needed to be equalized without creating downtime. The steward, needed in the front part of the plant outside the paint floor, had returned to his department via a different route at the other end of the conveyor in another department.

In the paint department end, the conveyor would run smoothly with no stops. In the other department, it would move in small repeated staccato jerks as it rounded a corner there. The perception was that the conveyor was moving slower than at the other end.

It took a good deal of explaining and use of stop watches and patient communication to finally settle the matter. Even as the steward walked away, he said he still felt that somehow management was pulling something. After using all my effort, I probably only obtained a draw in the matter.

Automating "burden"

There was the time when I was manager of a production area early in my career that a computer and its operator made life interesting.

The plant was in the midst of converting all the items in "Burden Stores" from manual to electronic records and invoicing. Our modern approach to management then said the use of the term "burden" was outdated, and from that time forward, we would use the term MRO (maintenance, repair, and operational needs). There also were new forms for ordering these supplies.

One of the requisitions was for toilet tissue. An order for so many sheets of paper was written as was always the case. That was the unit required whether buying a roll or a case of paper. Ordinarily, when the requisition was filled, the person pulling the order knew the system and gave us just a case. When the entry was made by a new computer keypuncher, however, it was entered as each sheet being a case. Our budget was debited for $100,000 worth of toilet paper, and the computer entered an order to the supplier (automatically) for the same. This actually happened. The only thing that kept us from being buried in TP and me being fired was our astute supplier who called to verify the order. Ultimately, everything was worked out.

The point here is "cover all your bases." When integrating new procedures and systems, brainstorm every contingency to backstop catastrophe.

The perils of empowerment

At certain times during the year, we went through large increases in schedule for painting school buses. Once again, the early use of computers and scheduling of materials had to go through a learning curve. School buses meant a certain yellow paint was required.

The schedule for buses would create a large demand for this color over and above normal because of seasonal considerations and demand. In those days, our computers were programmed to look at average use over the previous three months and release accordingly. This meant that it was quite likely we would run out of yellow paint in the first month of a schedule increase and have it running out of our ears the first month after the schedule went back down. No accommodation was made for cyclic demands. Peaks in demand were handled by manually intervening in the computer program.

Responsible for scheduling paint was a dear little lady who had manually scheduled paint for years before the change in technology. The first time we approached the schedule increase with the new technology on line, I assumed manual intervention would be required. I asked the lady if she had increased the yellow paint to meet our needs. She said she hadn't because we did not need any. When I pointed out the new schedule, she said that it did not tell her what color to order. I pointed out that that model of truck denoted school buses, and in this case, that meant yellow paint about 98% of the time, and we needed 1,000 gallons.

Ten days later, I got a call from our folks on the receiving dock who wanted to know where to put all the cases of paint they had received. What cases of paint? We bought paint by the drum, not cases. We had 1,000 gallons of school bus yellow in gallon cans, not drums.

True, I had told the scheduling lady 1,000 gallons of paint, but it was intended to be 20 drums of paint, each with 50 gallons of raw paint. We needed it in drums so we could add reducer and agitate it before adding it to our circulation systems. She literally gave what was said and did not question why we would not want drums anymore. Here the manual intervention went too far, as the unit size had been overridden.

Once again, we are reminded that communication can be a sticking point among workers and between workers and machines. (This story ended on a strange twist. One of the school bus body companies had missed the mark and came up short in material, called us for a help-out, and we came out heroes by not only having the paint, but having it in gallon cans as they normally used it.)

The bottom line

It took more than 5,950 years of recorded history for man to run a four minute mile. After Roger Bannister broke the barrier, it was broken 317 times in the following two years.

The same thing happens in the manufacturing world. Even if you break through and reach new levels of performance, it will be only a short time before competitors are matching that performance or beating it. As managers, we must constantly be improving performance.

In the following paragraphs is a collection of thoughts, data, and philosophies I have gathered over the years which have helped me understand and put perspective to numbers and situations in my personal and professional life. Such a perspective is necessary in today's business world. In our daily lives, we need to think logically and have the faith to rise above circumstances and not be bludgeoned by them.

One of the secrets to success is finding a skills niche and excelling in that niche. It is not necessary to be all things to all people or situations. I am reminded of a major paint company plant that has a two-element display in its lobby.

The first display is a huge cross-section of a redwood tree. On the rings of the tree are small signs that denote the size of the tree when most major events occurred from before the birth of Christ until the present. It is both interesting and humbling to realize that that tree stood during all that time.

The other exhibit is a small picture of a huge tree growing out of a tiny crack in a cliff of stone, standing straight and healthy against the stark rock outline. Captioned under the picture is a profound statement: "Grow where you are planted."

In my journey through my painting career, I was to be reminded many times how people rise from adversity to stand straight and healthy even though they came from difficult surroundings. One person–or even one achievement–can make a profound difference in our lives. Consider the following.

It takes the average person seven minutes to smoke a cigarette. If a person smokes 20 cigarettes a day for the 35 years of a working career, he or she will have used up 29,765 hours of time.

If each cigarette represented instead just one single idea, think how many problems could be solved, to say nothing of the money saved.

If you began to count dollar bills at the rate of one a second, it would take you 11.57 days to count a million dollars. At the same rate, it would take you 31.69 years to count to one billion dollars. For a trillion, it would take you 31,688.09 years. (The current national debt is four and a half times that much.)

Such number games bring home just how much money each of us is involved with in our daily business. They have helped me understand just how much a million dollars represents, for example, and I can now grasp the magnitude of what that amount of money can do in terms of both sales and costs. It drives home the need for constant vigilance in cost reduction and productivity. (To this day, however, I have never fully grasped what a billion dollars means.)

How goes the ship?

The following was excerpted from a performance review I authored nearly 20 years ago. It is still essentially up-to-date. It is included here verbatim.

Last week you requested an overview of our Division as seen through my eyes and work efforts. The following comments and observations are then offered.

The primary task ahead is to reestablish the #1 asset to success. That asset is to get ourselves "to want." This is a blunt statement, but so simple and true. Too many of our people no longer hunger or want to reach for a star. They are safely protected in the womb of mother and fed by the umbilical cord of the Company.

The time has long passed that management can hold a whip over employees and obtain results. Even the dangling carrot does not motivate as it once did. This fact is rather simple to recognize if you take the time to converse with employees. Most of them are really happier with their jobs than we realize. Being normal people, most do not assume risk well. After a few years on the job, they begin to find their slots and can remain there very satisfied.

Houdini has been quoted as saying, "There is no trick in pulling the rabbit out of a hat; the trick is putting it in." Management must find a way to put a rabbit in the hat for the future.

Too often the employees do not see anything in us to get excited about. We do not show our love for the work at hand nor do we believe in what we do. At all levels in the world, 20% of the people perform 80% of the improvements. Some of our planners would qualify for the job of cruise director for the Titanic.

We must lift people up to our level of excitement, not move down to their level of complacency. Our company is a living example of experience not being the best teacher. When you learn by our experiences, it is too costly. We cannot live long enough to learn just by experience. Once a cat sits on a hot stove, it does not live to sit on a cold stove. We have to learn by OPE (other peoples' experience) also.

(The major part of this review dealt with specific business situations and is omitted. The closing portion contained the following.)

If we think consumerism is a thorn in our side today, it will be a pistol to our back tomorrow. We will be using new washer products, new paints, and new substrates to counter these trends. Reduction of effluents will cause us to change our methods and coating materials. Marketing may have to revise sales identifiers on certain products due to lead and chromates being employed in those colors. We will have to spend capital monies in the next five years to produce product.

Education of new methods and materials will have to rank high in the next five years. School is never out for the professional! Unfortunately, the more I seem to know about a technical subject, the harder it is for the normal employee to understand. When it is simple for people at high levels after knowledge is acquired, it is very new and confusing to those who work for us. In every case, when I think something is easy, it is not. When that feeling hits me, I really begin to get concerned. It just works in reverse to the manager's ability so often. When the manager thinks it is easy, he fails to communicate to the others around them, and we are in over our necks.

It is not a gloomy future. It will require constant self-appraisal and work to make it happen. We shouldn't be too critical of ourselves in the process. Suicide is the highest form of self-appraisal. We shouldn't go that far.

Thank you for this opportunity to contribute my views. They probably are not the same as others' here. This is in itself good. If we had the same views, you wouldn't need me. It is out of such efforts we prove, test, and improve.

(End of report.)

The company changed very little. It did not attempt real change until it nearly went bankrupt. How goes our ship? It almost became the Titanic.

Recognition and self-esteem

There was a fellow who could not read or write. He tried very hard to earn a living, but it was difficult and results were meager. Many folks in the community had a soft spot for this fellow, although some made fun of him.

One of the community leaders was a bank president. He had grown up with the illiterate fellow and wanted to help him. The banker talked to him and convinced the fellow he needed a bank account to pay his bills, save for the future, and look toward his long-range needs. The ignorant fellow asked how he could do business since he did not read and write. The bank president said he would set things up with his tellers so all the poor fellow had to do was make two "XXs" and everything would be okay. This was done, and years went by with the man using this method of handling money.

Then the poor fellow won the lottery. He was a real celebrity. He was interviewed by the TV and press. He was instantly famous.

One day a teller brought to the attention of the bank president the fact that the illiterate, but now rich, fellow was using three Xs instead of his normal two in his transactions. Was this still okay, or what was happening? The bank president indicated he did not know what had happened, but he would find out. So he summoned his childhood friend and asked why he was now using three Xs instead of two.

The fellow replied, "All my life I have been a nobody. Now that my life has changed and I am a celebrity, it was my thought that perhaps I should now include my middle name."

Think about how ego and other things change about people who are suddenly given recognition.

Psychographics

Catching the ear of industry management of late is a concept called psychographics, the demographics of the mind. The following excerpts are taken from a presentation given as part of a training management seminar.

Demographics describes the physical attributes of societies, areas, and geographical divisions. Psychographics refers to the demographics of the mind. They help us understand what promotes behavioral differences.

Today we are encountering three distinct generations active in the industrial management climate. This interaction is altering the ability of these groups to participate in a cooperative manner toward directing and implementing the changes necessary to remain competitive today.

The three generations are quite far apart in some ways and differences among them are brought about by the cultural experiences of their members during their pre-employment years of development. Education and extended life expectancy have forced the interaction between the groups of people more than in the past. New management philosophies of team concepts and shared responsibility have thrust people together in both greater degrees of decision making and ability to contribute. Let us examine each of these groups.

Pre-boomer–fifties-sixties age group. This group came out of the Depression and WW-II periods. They focused on making do without and preparing for a better future. Hard work would ultimately bring rewards, and most had nowhere to go but up. Working and paying your dues in life would entitle you to a position of leadership and authority. Many worked or used the G.I. Bill of Rights to go to college.

The group had traditional values, sought stability and security for themselves and their families. They shared what they had in times of trouble and yielded where vested interests were involved to secure common goals.

Baby boomer–thirties-forties age group. Baby boomers are accustomed to having their own way. They have been catered to since birth and are loyal to themselves, not groups. Being a bit schizophrenic, they have a strong need to pursue what contributes to their personal status.

If they don't "own" a decision, they tend to not support programs coming out of those decisions. Their self-identity is so strong they are unwilling to give up or share involvement they feel would affect their own view of themselves. They want to satisfy desires now — at once — not pay their dues and earn rewards and authority levels. This group will move from job to job in an effort to find opportunities in companies where they can have an image of authority and satisfy their schizophrenic status contribution. This group is short-term oriented, compared to the pre-boomer, and is basically for sale in differing ways.

This generation saw grade schools built when they needed them, high schools provided, expanded colleges and easy ways to enter and finance education. The music business changed for them, fast food came along for them, dress codes, moral codes, and even automobile designs were created for their desires.

They expect the world to stand aside for them. When it does not, they will normally attack people around them as ignorant and out of touch. They will make every possible effort to change the conditions at any cost to obtain their goals. Without doing this, self-esteem would be badly compromised. When this cannot be done, they move on to another job to try and satisfy their insatiable wants.

Baby buster–twenties age group. The generation of baby busters coming into the employment scene today is the most educated, self-indulgent, and basically non-competitive group yet.

They feel entitled to everything at an early point in their life. There is little loyalty to the institutions of anything found in what we call established society. In many cases, this group feels all who went before them messed things up in the world. They feel there will not be enough to go around and to satisfy what they feel are their entitlements.

Theirs represents probably the most short-term thinking ever in our society. They save less money and tend to abuse credit. Most are more aware of the environmental issues, join causes, and believe government is letting them down. In a world that is not offering jobs and services to the degree known to the baby boomers, they want to make new rules. In fact, they tend to make the rules as they go.

While they tend to be better educated in terms of formal diplomas and degrees, many are not as skilled in basics as their forebears. Many can use computers, but cannot balance checkbooks, make change, cook, or make simple household and auto repairs.

Companies today are faced with building their work forces from these groups, more likely from pre-boomer and baby boomer groups than the youngest group.

Certain predictable disputes arise between these groups.

- Between executives and committees.
- Within committees.
- Among staff and committees.
- Between executives and staff.
- Within staff.

Symptoms are usually as follows:

- Inability to decide;
- Verbalized complaints about style or values;
- Personalization of disputes;
- Battles among veterans and new members of programs, power;
- Behavior among work groups that reflects frustration, anger, resentment, abandonment;
- Continuous attempts to change rules or policy;
- Shifts of loyalty and identification to appease factions.

How to sell a program

I see selling a program as containing at least three elements: *educate, motivate, and activate*!

1. Educate your constituency. Give to people and treat them as intelligent human beings.
2. Motivate them, and give them reasons to support you.
3. Activate them so they do the front-running for you in all the important activity areas.

These three key points are known as unique selling points (USP). Direct all activity to these. Create graphs and other visuals reflecting these efforts to leave "mind pictures" of the idea or program.

Keep the outline content simple, and ensure its accuracy. Provide for both a feel for the speed of opportunity and a response opportunity for objections.

When you own something, you normally treat it differently. This is true of ideas or property. In your business world, it may only be ideas, but the theory is the same. The more people *buy into* an idea, the more attention that idea will enjoy.

How goes the ship? — Part II

This is a follow-up report to the original How goes the ship? performance report appearing earlier in this chapter. It is still quite current in its message and its applicability to most paint operations. The efficacy of its message was borne out when 10 years later I was rewarded with a huge pay raise for having the courage to warn executive management of great risk.

"This is the way we are going to do it!" A brief but powerful moment occurs in life that transforms a person into a manager. It is the realization that they have just made a decision. "This is the way...!"

Today, management is searching frantically for gigantic breakthroughs in performance. Certainly there is a place for these. Unfortunately, the real truth is we had a weak system to begin with.

I would suggest we are too often concerned with activities. Results from simple acts of leadership done promptly are the identifiable signs of good management. Results project the truest image.

When a manager pronounces the words and acts in a manner that translates this simple message to employees, that manager has a results-oriented record.

Such an environment does not emerge in an organization where fear dulls incentive. It does not come in an organization where ignorance has not been replaced with the confidence of knowledge. It does not come in an organization where personal integrity is no longer a prerequisite to qualification.

Are company policies symbolic or fact? Cleverness today is often substituted for loyalty by management. When put to a test, an ounce of loyalty to your employers is worth a pound of cleverness. Not blind loyalty, mind you, but the old-fashioned kind. The kind where one is willing to risk his or her reputation or incur a little momentary wrath from the boss. It is not improper to warn the boss when something is "tilting" in the organization. It is risk assumption.

Everything listed here is "an act of leadership." Each is a little brick used to build an edifice of lasting value. Each in itself is not very impressive but so valuable to the end result.

We have people who would construct our edifice in another manner. They are too smart to be makers of bricks. Their method is to get a block of granite large enough to meet the dimensions. They then would chisel out the rooms inside to provide the finished product. In the end, they have proven how smart they were.

It doesn't matter if there was a little material handling problem with the block of granite. The scrap rock that was removed and hauled away didn't count because we had good housekeeping; it didn't pile up, and it wasn't noticed. Besides, we sold the trucks to the haulers, so it was "good business" on our part, wasn't it?

The accomplishment is lost in the expense of investment, expensive material handling, scrap, labor, and public relations. These monuments to success soon become granite tombstones for posterity.

We shall not be able to leave these kinds of tangible monuments for history to ponder over centuries from now. The pyramids were houses for the dead. They could remain unchanged. Our house cannot.

What we build will be changing, because it is a growing and living entity. Our results will pass to future generations in different form because of this change and the vitality it generates.

Now we have come back full circle to our "act of leadership." Management at all levels must first say, "This is the way we are going to do it."

Middle and front-line supervision must then implement it, but top management will always be expected to say it first.

Being on the front end makes the difference between executive and other management.

Our organization needs more "I.D." in this recipe. This is our weakest segment. Everybody plans, a few refine, and all will wish to enjoy the results. Instituting and Demonstrating is where we must really shine.

- **P**lan your goals
- **R**efine your ideas
- **I**nstitute them
- **D**emonstrate ability
- **E**njoy the results

Good management here must make the transition from idea level to actual product.

This year you will see a concerted effort to make this middle and front-line supervision more capable of establishing their identity in this plan. You can then build any plan you wish in our finishing systems.

On human anxiety

Early in my career, I took note of an interesting phenomenon: nearly every time some great improvement took place, workers would first exhibit euphoria over the achievement, then slowly their attitudes would begin to almost reverse.

They were suddenly becoming afraid that their jobs would be eliminated if all the problems were solved. To overcome this feeling, I had to find some way to make them feel needed again.

To avoid people from sabotaging product or finding ways to create problems, I would create some of my own. None that would affect daily production, but ones focusing on trials of new materials. These new materials were normally screened by me at the suppliers before shipment to our shop. It was a good training exercise to occasionally bring in a product which was known to have some shortcomings.

The testing and evaluation of this type product would always satisfy workers' need to pinpoint a problem. It both reinforced their sense of security, and it was good training.

It is necessary to have a means of employing people in new production at about the same rate improvement is being made. We will then see more product created with the same amount of labor. It takes strong leadership, hard work, and careful planning to avoid laying off people.

Vaulting the stone wall

How many times have you drafted a plan and presented it, only to have someone up the ladder give it a cursory look and dismiss it with a remark like, "We don't have money for such a program"? Too many times good ideas have been stonewalled without consideration of what the payback might be. This is especially true when interest rates are very high: financial managers would say, "We can't afford to pay 18% interest."

Once again, we are not paid for having brains, we are paid for using them. Any 12-year-old can be taught to say, "We can't afford to pay 18% interest." A real financial planner will recognize that even in the tightest of money markets, there may be programs that will justify borrowing money if the payback is right. I have had this happen to me when there were paybacks within three months.

You don't give up in these cases. It will for sure be a hard sell to convince the decision maker that paying 18% interest is in the best interest of the company.

Other management tidbits I have picked up during my career follow. Some — maybe all — might be helpful to you.

- Many companies have a "Kill the messenger mentality" toward bad news. It isn't long before no messages percolate upward. If promotions were frozen, with most being held to three to five years, people would accept the messages and be more responsible for actions taken long term.
- Does your company have a good performance evaluation system? It is not unusual to find everyone performing well by individual evaluations, but the total company enterprise could be losing.
- Somewhere in my travels in New England, I saw the following quote on a school building:

    Goodness without knowledge is weak and feeble.

    Knowledge without goodness is dangerous.
- Some people aren't good for you. They are "toxic." If they do not encourage you to learn, grow, stretch, and then empower you, stay away from them. If you associate with losers, you will probably become one.
- A sandwich is a classic example of a well designed product. It can be produced many ways to satisfy a need, it handles well, and has no dunnage to dispose of when finished.
- Remember the things that keep you going, to get up off the floor, and prove people wrong. Buster Douglas won the world championship against a lot of odds for "Momma." Then with $24 million guaranteed, he lost the incentive.
- Things change. Our pioneer forefathers crossed the untamed wilderness, fighting and paying their way as they went. Now their descendants won't even drive to town without a credit card.
- Never let anything take away your dreams. Dreams are the fuel of a desire to improve and succeed.

Bibliography

Anthony, Robert N.; *Management Accounting*, Third Ed., Richard D. Irwin, Inc,; Homewood, Ill. 1963.

Staff, Binks Manufacturing Co.; "Dirt in Finish"; presentation, *Finishing '91* conference, Society of Manufacturing Engineers; Dearborn, Mich., October 1991.

Staff, Bureau of Business Practice; *ISO 9000 Handbook of Quality Standards and Compliance;* Prentice Hall; Englewood Cliffs, N.J., 1992.

Carroll, P.B.; "The Failure of Central Planning at IBM"; *Business Week*; October 21, 1991.

Chaffman, B.M., and Talbott, J.; "Activity Based Costing in a Service Organization"; *CMA Magazine;* December/January 1991.

Staff, Deloitte & Touche; CAM-1 ABM Work Group Expert Meeting; presentation by Deloitte & Touche, May 1992.

Deming, Dr. W.E.; *Out of the Crisis;* Center for Advanced Engineering Study, Massachusetts Institute of Technology; Cambridge, Mass.; 1986.

Drucker, P. F.; "The Practice of Management"; Harper & Brothers.; New York; 1954.

Dudley, R.P.; "Ten Most Common Mistakes in Wastewater Treatment"; technical presentation, *Finishing '91*; Association For Finishing Processes, Society of Manufacturing Engineers; Dearborn, Mich.; October 1991.

Goodman, E.; "Just Plain Overwork Alienates Employees"; *Washington Post; Washington Post* Writers Group; Washington, D.C.; 1992.

Grear, R.D.; "Liquid Paint Defects"; tutorial presentation, Society of Manufacturing Engineers; Dearborn, Mich.; 1992.

Grear, R.D.; "Paint Facilities Environmental Policies"; technical presentation, *Automation of Paint Lines*; Society of Manufacturing Engineers; Dearborn, Mich.; 1985.

Grear, R.D.; "Pre-Startup Checklist for Liquid Spray Lines"; Association for Finishing Processes, Society of Manufacturing Engineers; Dearborn, Mich.; 1991.

Kepner, C.H., and Tregoe, B.B.; *The Rational Manager*; Kepner-Tregoe, Inc.; Princeton, N.J.; 1976.

Laich, D.; "Entitlement"; *Navistar Update*; Navistar Corp.; September 1992.

Looney, R.; "U.S. Quality Slurs Don't Jibe"; *Las Vegas Sun;* Las Vegas, Nev.; 1992.

McLeod, R.G.; "Whining or Winning"; *San Francisco Chronicle;* San Francisco, Calif.; 1992.

Price, J.L.; "Manufacturing Matters"; *Manufacturing Engineering;* Society of Manufacturing Engineers; Dearborn, Mich.; February 1992.

Reseland, J.C.; "Troubleshooting Spray Washer Systems"; technical presentation, *Automation of Paint Lines*; Society of Manufacturing Engineers; Dearborn, Mich.; 1991.

Richards, M.D., and Nielander, W.A.; *Readings in Management*; Second Ed.; Southwestern Publishing; Chicago, Ill.; 1963.

Index

A

Accountability, 10
Accounting practices, 221
Accounting systems, 67, 70, 73, 74
Activity-based Costing (ABC), 81, 82
Activity-based Management (ABM), 81, 133
Air pollution, 204), 205, 210, 212, 215
Air Pollution Federal Requirements, 213
Air toxics, 202, 213
Air-dry technology, 99
Alkyds, 99
American Society for Quality Control (ASQC), 131
ANSI-ASQC Q90, 129
Application equipment, 111
Application techniques, 25
　electrostatic, 35, 157
Aqueous cleaners, 97
Assets
　creators of, 37
　fixed, 77
　users of, 37
Authority, 10, 23, 66, 198
　financial, 9
　levels, 9, 34
Automatic transport vehicles (ATVs), 110
Autophoretic systems, 97, 99, 119

B

Behavior, 23
　corporate, 17
　patterns, 35
Benchmark, 45, 128, 147, 165
Biochemical Oxygen Demand (BOD), 208
Booths, 111, 219
　paint, 14, 112, 166, 167, 198
　spray, 159, 191
Brainstorming, 169

Budget process, 78
Budgets, 65, 66, 77
　architecture of, 78
　capital, 77
　control, 90
　goals, 78
　operating, 77, 78
　planning, 77, 90
　program, 78
　"responsibility," 78
　time period for, 78
Bulk materials, 73
Burden, 74, 88
Business,
　plan, 68, 72, 75
　strategies, 62, 65

C

Capital investment, 91, 102
Change, 57, 165
　regulatory, 16
Checklist
　pre-startup, 184
　safety, 220
　startup, 199
　troubleshooting, 120
Chemical Abstract Service Number (CAS or CASN), 203
Chemical conversion coatings, 95
Chemical Oxygen Demand (COD), 208
Chlorinated degreasers, 97
Chrome conversion process, 95
Clean Air Act (CAA), 98, 202, 210
Clean Air Act Amendments (CAAA) of 1990, 211, 212, 213, 214
Clean Air Program (CAP), 214
Clean Water Act (CWA), 207
Cleaning products, 112
Clean-room, 178
　facilities, 109, 177
Coatings, 91, 211, 215, 223
　liquid, 94

261

organic, 119
powder, 92, 94
thermoplastic, 93, 94
thermosetting, 93, 94
Co-generation, 112
Collaboration, 198
Commitment, 34
　management, 19, 34
　corporate, 19
Common denominator, 53, 65
Common sense, 34
Communicate, 46
Communicating
　policy, 17
Communication, 10, 18, 19, 22, 68, 71, 75
Communications, 69, 227, 231
　electronic, 79
　skills, 24
　techniques, 64
Competitiveness, 16, 26
Comprehensive Environmental Response, Compensation and Liability Act (CERCLA), 208
Compressed air supplies, 191
Conductivity, 92
Confidence, 22
Conformist people, 36
Consultants, 168, 173
Contaminants, 12, 106, 182, 214
Contamination, 23
　silicone, 166
Contingency management, 167
Continual harmony, 227
Continuity, 37, 57
Continuous improvement, 9, 63, 72, 77, 81, 120, 127, 133, 166, 168
Contractual constraints, 183
"Contrarian" approach, 108, 109
Control, 66
Controls, 113
Controlling, 6
Conveyors, 13, 14, 183
Corrosion, 13, 22, 203
Corrosion protection, 92, 95, 96, 119
Corrosion resistance, 93, 97
Corrosive hazardous wastes, 206

Cost, 11, 13
Cost-control, 113, 118
Cost drivers, 114
Cost-efficiency ratio, 114
Costs, 115
　fixed, 67, 80
　variable, 67, 80
Creativity, 45
Cross-linked thermosetting resins, 100
Cure volatiles, 216
Curing, 25, 92, 94, 95, 99, 101, 115, 220
　films, 100
　processes, 100

D

Data gathering
　techniques, 64
Decals, 102
Decision making, 163, 169, 170, 172, 173, 174, 175
　consensus, 10
　joint, 170
　team, 10, 170
Dedication, 59
Degreasing, 96
Deming, Dr. W. Edwards, 126, 127, 131, 133
Department of Transportation, 205
Depreciation, 80
Design changes, 144
Dip systems, 119
Dipping, 99
Direct labor, 67, 70, 79, 88
Direct materials, 67, 72, 73, 88, 221
Dirt, 106, 177
Discipline, 59
Discoloration, 100
Diversity, 45
Downsizing, 41, 43, 46, 165

E

Education, 75
　formal, 71
　on-the-job, 71
Ego-driven people, 37

Electrocoat (e-coat), 97, 99, 119, 228
Electronic data interchange, 222
Electronic monitoring, 168
Electronic surveillance, 126
Electrostatic systems, 119, 157
Emergency Planning and Community Right-to-Know Act (EPCRA), 208
Emissions, 93, 158, 159, 201, 210
  calculations, 160
Employee Right-to-Know Law, 208, 209
Empower, 24
Empowered, 9
Enamel, 228
Energy, 111, 112
Engineering, 144
Enterprise integration, 222
Entitlement, 41, 42, 44, 45, 46, 72
Environmental
  compatibility, 11
  compliance, 26, 94, 110, 113, 116
  considerations, 106, 111, 201
  laws, 26
  mandates, 25
  regulations, 97, 107, 202, 205, 207, 208, 209, 214
Environmental Protection Agency (EPA), 98, 201, 207, 210, 211, 212, 214, 216
Equipment, 8
Ergonomics, 126
Esprit de corps, 50
Ethics, 225
European community (EC), 129, 130
Evolutionary
  approach, 63
  change, 16, 26
  improvements, 20
  management, 14, 91
  operation, 57
Exempt solvents, 201
Experience, 41, 144, 168

F

Facilities, 68, 108, 111, 198, 199
  clean room, 109, 177

Failure Modes and Effects Analysis, 141
"Farm" systems, 49
Fiberglass, 92
Fiduciary responsibility, 66
Filters, 166
Filtration, 111
Finishing, 30
  as a strategy, 11
  development of operation, 14
  fit, 30
  human requirements, 14
  operations, 25
  philosophy, 11
  physical elements, 14
  planning processes, 13
Fishbone chart, 147, 148
Fisheye cratering, 166, 228
Flash-off, 115, 211
  zones, 105
Flexible, 46, 127, 133
  operation 57
Flexibility, 16, 20, 56, 63, 80, 91, 94, 95, 194
Flexible substrates, 98
Flow coating, 99
Fluidized powder systems, 99
Force-dry technology, 99
Forecasting, 65, 70
Formal education, 71
Freons, 205

G

General foreman, 30
Generic resins, 99
Goals, 11, 64, 68
  long-term, 43
  short-term, 9
Good housekeeping practices, 219
Guns
  air-atomized, 14
  airless spray, 14
  electrostatic, 13, 14, 25, 167
  HVLP, 14, 26, 98, 99, 119

H

Hanging, 95
Hazard Communication Standard, 208
Hazardous wastes, 205, 206, 207
  corrosive, 206
  ignitable, 206
  reactive, 206
  toxic, 206
Health, 200
Health considerations, 201, 217, 220
High solids, 153
Housekeeping, 31, 56, 182, 186
Human resources, 8, 19, 70, 168, 178, 224

I

Ideology, 118
Ignitable hazardous wastes, 206
Image, 72, 182
Implementation
  evolutionary, 14
  revolutionary, 14
Improving a process, 166
Indirect materials, 67, 72, 73, 88, 221
Industrial engineering, 30
Information technology, 29
Ingenuity, 43, 133
Inspection work card, 144
Integrity, 17, 20, 34, 75, 79, 90
International Organization for Standardization (ISO), 129
Intuition, 168
Investment, 67, 72
Iron phosphate, 95
Islands of technology, 223
ISO 9000, 66, 125, 128, 129, 130, 131, 133, 143, 147
Isocyanate hardeners, 229

J

JIT, 111, 222
Job enrichment, 35, 42, 46
Juran, Joseph, 126

K

Kepner-Tregoe training system, 169, 170
Kickbacks, 222

L

Labor, 43, 75
  direct, 67, 70, 79, 88
  incentive piecework, 70
  indirect, 67, 70
  leadership, 231
  support, 70
Lacquers, 99, 228
Leadership, 5, 9, 10, 16, 20, 25, 26, 50, 51, 57
  needs, 58
  strategies, 45

Liquid technologies, 110
Liquids, 101
Loss of gloss, 100
Low solids, 153
Loyalty, 34, 43, 108, 199
Lubricants, 12, 22, 92

M

"M & M" recipe, 18
Maintenance, 31, 70, 89, 104, 117, 120, 177, 182, 183, 223
  predictive, 114, 168, 199
  preventive, 114, 168, 197, 199, 223
Maintenance deep clean, 184
Maintenance facilities, 224
Maintenance organization, 199
Maintenance procedures, 178
Malcolm Baldrige National Quality Award, 141, 149
Management, 5, 7, 26, 51, 57, 227, 230, 231
  career movement, 17
  "Charlie Tuna," 44
  charts, 31

evolutionary, 91
low-visibility, 26
needs, 57
participation, 17
philosophy, 20, 43
practice, 19
proactive-preventive, 164
science, 19
staff mix, 31
strategies, 45
technical, 34
techniques, 43
Management's role, 224
Manager
  area, 30
  general, 30
  operational, 30
Managers
  results-oriented, 56
  task-oriented, 56
Manipulative people, 36
Manufacturing engineering, 30
Material handling, 12, 13, 31, 57, 96, 111
Material resources, 224
Material safety data sheets (MSDS), 205, 208
Material transportation system, 95
Materials, 8, 25, 30, 43, 44, 73, 75, 109, 111, 203, 207, 224, 229
Maximum Achievable Control Technology (MACT), 212
"Me" syndrome, 42
Measurement, 143
Measurement system, 143, 146
Measuring quality, 128
Media blasting, 95
Mergers, 43
Metal substrates, 100
Metals, 94
  galvanized, 93
  hexavalent chrome, 95
  pre-coated, 93
  uncoated, 93
  zinc-clad, 93

Morale, 10, 69, 70
Motivation, 9
MSDA sheets, 223
Multi-toning, 102, 105
Murphy's Law, 68

N

Needs, 46
Negative behavior, 230
Negative communications, 227, 231, 232
Nitrogen oxides (NOx), 203, 204
Nonattainment areas, 213
Nonconductive substrate, 92
Nonmetal substrates, 93, 100
Nonvolatiles, 215
Now Generation, 42
NPDES, 207, 212

O

Objectives, 62
  short-term, 9
Occupational Safety and Health Administration (OSHA), 205, 208
Old-school managers, 170
Operating procedures, 231
Orange-peel, 93, 100
Organizational
  charts, 28
  philosophy, 89
  structure, 18, 20,
Organizing, 6
OSHA safety rules, 109
Outsourcing, 224
Oven curing time, 117
Ovens, 112, 191, 211
  convection, 100
  infrared, 100
Overhead, 67, 74, 75, 79
  fixed, 74
  variable, 74
Overoptimism, 64
Overtime, 183

265

Overtime fraud, 230
Ozone, 202, 203, 204, 209
Ozone depleting chemicals, 213
Ozone depletion, 204

P

P&M, 88
Paint booths, 112, 166, 167, 198
Paint defects, 169
Paint facilities, 102
Paint mileage, 154, 155, 156
Paint ovens, 220
Paint sludges, 207
Paint solids, 151
Painters' math, 151
   formulas, 161
Painting "air," 100, 119
Painting strategies, 91
Paperless communication, 222
Partners, 198
Partnership, 197, 198, 199
Paternalistic, 127
Patience, 22
Payback time, 67
Performance, 191
   economic, 7, 8
Plan
   business, 62, 63, 64, 65
Planning, 6, 69
   budget, 77
   business, 61, 66
   long-term, 113, 127
   strategic, 61, 63
Planning facilities, 91
Plastic, 92
Policies, 9, 64, 69, 178
Powder coating, 92, 94
Powder technologies, 110
Powders, 97, 101, 119
Practical plant capacity (PPC), 104
Pressure pots, 110, 228
Pre-startup checklist, 184
Pretreatment, 13, 22, 25, 56, 88, 89, 92, 95, 112, 117, 119, 184, 186
Pretreatment chemicals, 197
Pretreatment maintenance, 186
Prevention, 164
Pride, 141, 149
Primer/sealers, 97
Primerless systems, 101
Primers, 92, 94, 96, 97, 112
   self-etching, 98
Priorities, 68
Proactive-preventive-management, 164
Problem analysis, 171
Problem solving, 163, 164, 166, 167, 170, 173, 174
Procedures, 64, 69, 143, 149
   maintenance, 178
   standard operating, 66, 143, 144
Process, 11, 114
   decision-making, 43
   engineering, 25
   monitoring, 135
   routings, 30
   variables, 144
Processes, 8
Product
   certification, 130
   design, 12, 13
   engineering, 30
   integrity, 66
   liability, 130
   quality, 66
   team, 222
Production flows, 64
   forecasts, 159
   manager, 31, 89
Productivity, 13, 19, 28, 43, 72, 106, 114, 116, 118, 119, 133, 182, 231
   increases, 127
Profitability, 118
Profits
   short-term, 43
Property, 8
Protective coatings, 217
Purchasing, 30
Purchasing policies, 221

Q

Q-Circles, 131
Quality, 11, 13, 19, 25, 26, 56, 95, 96, 100, 101, 109, 116, 120, 129, 133, 141, 146, 147, 165, 177, 178, 191, 221, 222, 223, 224
  control, 30, 125, 134, 149
  first-time, 44, 219
  Management and Quality Assurance Standards, 129
  of information, (QOI) 124
  standard, 140
  system registration, 130

R

Racking, 92, 94, 95
Reactive hazardous wastes, 206
Reactive management, 200
Reactive problem solving, 164
Reasonably Available Control Technologies (RACT), 210, 211, 216
Recruiting, 50
Registrar Accreditation Board (RAB), 131
Regulatory changes, 16
Regulatory compliance, 200
Reliability, 191
Reorganizing, 27, 51
Resin systems
  solvent-based, 97
  water-based, 97
Resins, 201, 228
Resource Conservation and Recovery Act (RCRA), 205, 206, 208
Resources
  human, 8, 19, 70, 168
  material, 8, 19
Restructuring, 41, 43
Retraining, 167
Reward, 10
Right-to-know laws, 208
  Comprehensive Environmental Response, Compensation and Liability Act (CERCLA), 208
  Emergency Planning and Community Right-to-Know Act (EPCRA), 208
  Employee Right-to-Know Law, 208
  Hazard Communication Standard, 208
  Superfund Amendments and Reauthorization Act (SARA), 208
Risk, 6
Robotic application, 14
Robots, 63, 183
Rules
  work, 16

S

Sabotage, 227, 229, 232
Safety, 11, 31, 70, 114, 177, 186, 198, 200
  OSHA, 109
Safety checklist, 220
Safety considerations, 201, 217, 220
Scheduling, 13
Science, 7
Scientific management, 126
Sealers, 112
Security, 46
Self-etching primers, 98
Self-improvement, 72
Seniority, 49, 50, 167, 183
Sheet molding compound (SMC), 94
Short-term benefits, 231
Shutdown procedure, 199
Silicone compounds, 228
Silicone contamination, 166
Simple degreasing, 95
Simulation, 69
Single sourcing, 223
Siphon cups, 110
"Skyhooks," 51
Sludges, 95
Smog, 203
Society of Manufacturing Engineers, 72, 169
Sociocentric people, 36
Solids, 152, 153
  high, 153
  low, 153
Solvent pop, 100

Solvent-based liquid systems, 110
Solvent-containing materials, 205
Solvents, 152, 158, 159, 160, 202, 203, 205, 208, 214, 219
  ignitable, 207
  listed, 207
  toxic, 207
Spangle, 93
Spreadsheet analysis, 126
Stability, 45
Staffing, 6
Standard operating procedures, 66, 143
Standards, 129
Startup, 184
Startup checklist, 199
Statistical process control, (SPC) 66, 125, 133, 143, 147, 197, 199
  charts 144
Statistical quality control (SQC), 126
Strategies
  business, 45, 62, 65, 69
  painting, 91
Striping, 102
Substrates, 25, 93, 94, 95, 96, 97, 100
  flexible, 98
  metal, 100
  nonconductive, 92
  nonmetal, 93, 100
Superfund Amendments and Reauthorization Act (SARA), 208
Superintendent, 30
Supplier relations, 221, 222, 225
Suppliers, 17, 25, 62, 63, 72, 143, 144, 168, 173, 197, 198, 199, 203, 223, 225

T

Takeovers, 41, 43
Taylor, 133
Taylorism, 28, 67, 126, 147
Tayloristic, 29
Teams, 27
Technical
  leader, 120
  supervisor, 31
Theft, 228, 229, 230
Thermoplastic, 93, 94, 100
Thermosetting, 93, 94, 100

"Think" games, 69
Third-party registration, 131
Threshold Limit Values, (TLV) 219
Time, 68, 69
Time and motion studies, 126
Timetables, 68
Topcoats, 94, 112
"Total costs," 115
Toxic hazardous wastes, 206
Toxics, 207, 220
Tracking, 66, 67
Traditional people, 23
Training, 22, 24, 25, 26, 67, 70, 71, 72, 75, 135, 144, 170, 182
  self-training, 71
  structured training, 71
Transfer efficiency (TE), 99, 115, 119, 154, 156, 157
  determining, 156
Tribalistic people, 23
Troubleshooting checklist, 120
Turnaround, 67

U

Ultraviolet (UV) rays, 205, 219
Uncoated metals, 93
Uncured paint films, 100
Union environment, 49
U.S. Labor Department, 43
Utopia, 227

V

Value
  -added labor, 67
  adding, 11, 56
  competitive advantage of, 11
Values, 49, 50, 51, 56, 227
Variable costs, 67
Vision, 26
VOC equations, 215, 216, 217, 218
Volatile organic compounds (VOCs), 13, 26, 73, 111, 151, 152, 158, 159, 201, 204, 205, 207, 209, 210, 211
Volatiles, 152, 215
Volume solids, 152

W

Wants, 46
Waste, 113, 115
  disposal, 191
  elimination, 114
  management, 115
  minimization, 98, 99, 114
  prevention, 25
  treatment, 25, 165, 182
Wastewater, 191, 197, 205, 207, 212
  discharges, 203
  treatment, 95, 191, 205
Water contamination, 203
Water treatment, 191
Water-based liquid systems, 110
Waterborne coating, 25
Waterborne paint system, 105
Work ethic, 41, 42, 43, 47, 59
Working conditions, 70
Wraparound, 157

Z

Zinc phosphate, 95
Zinc phosphate system, 105